Smarter
Tomorrow

Smarter Tomorrow

How 15 Minutes of Neurohacking a Day Can Help You Work Better, Think Faster, and Get More Done

Elizabeth R. Ricker

Little, Brown Spark
New York Boston London

Little, Brown Spark
Hachette Book Group
1290 Avenue of the Americas, New York, NY 10104
littlebrownspark.com

First Edition: August 2021

Little, Brown Spark is an imprint of Little, Brown and Company, a division of Hachette Book Group, Inc. The Little, Brown Spark name and logo are trademarks of Hachette Book Group, Inc.

The publisher is not responsible for websites (or their content) that are not owned by the publisher.

The Hachette Speakers Bureau provides a wide range of authors for speaking events. To find out more, go to hachettespeakersbureau.com or call (866) 376-6591.

Illustration credits can be found in the Notes at the back of the book.

ISBN 978-0-316-53515-1 (hc); 978-0-316-36651-9 (international edition)
LCCN 2021931988

Printing 1, 2021

LSC-C

Printed in the United States of America

To my family—Varun, Mom and Dad, Lindsey, Latha and GP,
and our newest member, Munchkin—I love each
of you and all of you so much.

To those of us who, at one point or another, worry that we must
choose between being true to ourselves and being successful
in the world as it is, I hope this book provides tools to
emerge from that struggle at once victorious and at peace.

Contents

■■■■ PART I ■■■■
Prepare to Neurohack

■■■■ PART II ■■■■
Find Your (Mental) Focus

■■■■■ PART III ■■■■■
Select a (Foundational) Intervention

■■■■■ PART IV ■■■■■
Select an (Advanced) Intervention

■■■■■ PART V ■■■■■
Train and Reflect

Author's Note

The material in this book is for informational purposes only. It does not constitute, and is not intended to substitute for, medical advice or to diagnose, treat, cure, or prevent any health problem or condition. No action should be taken solely on the information in this book. Always consult your physician or qualified healthcare professional on any matters regarding your health and before adopting any suggestions or following any recommendations in this book or drawing inferences from it. The author and publisher specifically disclaim all responsibility for any liability, loss, injury, damage, or other adverse effects that may result, directly or indirectly, from the use or application of information in this book.

Mention of specific companies, organizations, or authorities in this book does not imply endorsement by the author or publisher.

Where dialogue appears, the intention was to re-create the essence of conversations rather than verbatim quotes. This is a work of nonfiction. Dialogue was, in some cases, created for the purposes of entertainment. Names and identifying characteristics of some individuals have been changed.

Facts Matter: A Note on Notes

Throughout the book, you'll see little numbers perched above the words in many, many sentences. These numbers indicate that there was a specific source, usually a peer-reviewed scholarly research paper, that I was referring to when I made a particular statement. If you're curious about the source, just flip to the Notes section at the end of the book. There you'll find the sources listed in the order they appear, organized by chapter. Feel free to look up these sources and use the index; these are great ways to further explore any topic in the book.

Wherever you see the word "percent," I'm referring to something called an "effect size." This describes the effect that a treatment had on people's average cognitive performance. There are many ways to describe these effects, but I use percentile standing. For more info, check out the endnotes.[1]

A big thanks is owed to my science fact-checkers and beta readers. This team of more than a dozen neuroscience graduate students, professors, undergraduates, and research professionals hailed from Google, Harvard, Johns Hopkins, McGill, New York University, uOttawa Brain and Mind Research Institute, Stanford, University of California–Berkeley, University of Chicago, University of California–Davis, and University of California–Los Angeles. They pored over the manuscript and hunted for mistakes. If you see facts in the book that are wrong, it's probably because I didn't heed good advice from one of them. If something looks off to you, just shoot me a message at ericker.com.

Smarter Tomorrow

Introduction &
How to Use This Book

"The future is already here—it's just not evenly distributed."
—William Gibson

Time Investment: 11 minutes

Goal: To understand what you will and won't get out of reading this book

One question has driven me from city to city. It has pushed me to hunt down experts across the world. It has motivated me to scour hundreds of research papers. It has compelled me to jump from working in neuroscience research to technology startups and back to research—but research of a far more personal type. Finally, it has fueled my testing of dozens—no, let's be honest, hundreds—of apps, wearables, devices, and substances on myself. If I could answer this question for myself, I figured, I could answer it for others, too:

How can I upgrade my brain?

Not how can my doctor upgrade it. Not my teacher. Not my boss. Not my family or friends. How can *I* do it?

What does a "brain upgrade" mean? For me, it means getting consistent access to the best version of myself, not the one who slouches up on an average day. I want the version who learns fast, who recalls details that others forget, who juggles daily responsibilities without dropping anything important, who says just the right thing to make a friend feel better. The version of me who is bold, reliable, kind—and gets shit done.

Science fiction shows us what a dramatic mental upgrade could look like. With the help of a pill, a man becomes a financial wizard in weeks. With neural implants, a woman masters helicopter piloting in seconds. Other stories describe flawless memory, effortless language learning, and boundless creativity.

What if it didn't have to remain fiction, though? Even everyday upgrades could be life-altering: What if we didn't forget someone's name 10 seconds after meeting them, didn't snap at our significant others when they interrupted us (sorry, honey!), didn't click uncontrollably from silly cat video to silly cat video when faced with a deadline? What if we displayed monk-like calm when our siblings or toxic coworkers tried to provoke us? In short, what if we could tap into our better, smarter selves, not just some of the time... but most of the time?

Back to the Beginning

Did your school report card include phrases like "seems bright but is not working to her full potential"? If so, my elementary school days might sound familiar to you. One day, my teacher made the following announcement to the class: "Elizabeth is going to a special tutor because she does not know how to read like the rest of you." My cheeks burning, I toyed with the idea of running away rather than climbing the steps to the reading tutor's office. After a few minutes with Ms. Lecto,[1] however, two things became clear. First, she didn't think I was a hopeless case. Second, she both challenged me and trusted me to progress at my own pace. With my previous teachers, I had daydreamed and doodled constantly. With Ms. Lecto, my brain finally turned on.

With her help, I went from being one of the worst readers in my grade midway through the year to fully caught up by summer break. By the following year, I was reading and writing above grade level. A few years later, I was nearly unrecognizable. Instead of being the kid who doodled through class every day, I was earning top grades and winning writing competitions. Later, I earned one degree from MIT and another from Harvard.

What happened? How *exactly* had the transformation occurred?

Perhaps not surprisingly, I didn't encounter a lot of late readers among my classmates in graduate school. But I did discover something else there: research on reading trajectories.[2] It turned out that students who fell as far behind as I had tended to stay behind when they became adolescents, especially in the United States. Furthermore, delayed readers were far less likely to graduate from high school.[3] I had beaten the odds—but how? Other research papers revealed that for some children who would later struggle with reading, certain brain regions involved in language were smaller or under-activated, even before the children attempted to read.[4] I wondered whether my struggles could have been predicted, too. Or, was it that I simply wasn't paying attention? I did daydream a lot, after all. A Duke study tracked nearly 400 children from kindergarten through fifth grade and found a strong role for attention in predicting reading struggles.[5]

Because the brain changes so much over time and I never got a childhood brain scan or an attentional assessment, I'll never know whether my reading struggles could have been predicted. However I got there, it was clear that I later won the academic lottery. Ms. Lecto's personalized approach set me on a far happier academic trajectory than my homeroom teacher could have predicted. If attention was my problem, Ms. Lecto helped with that beautifully. In this book, you will learn about one of the key steps in upgrading the brain: discovering bottlenecks. The flow of water through a bottle is limited by the rate at which it can pass through the narrowest part of the bottle: its neck. In my case, my inability to control my attention was at least one of the bottlenecks preventing me from learning in the regular classroom. For you, the bottleneck may be something else.

Not everyone gets a Ms. Lecto. Frankly, I felt a little guilty at how lucky I had been. Being able to control and manage your mental performance is too critical to leave up to luck. Sometime between high school and college, I began to dream about brain upgrades more generally—not just for improving reading ability or improving attention. Were there evidence-based tools that could provide mental upgrades to anyone, regardless of their bottleneck?

For the next decade (and more), I pursued this question in a range of stressful, exciting, and occasionally comical conditions. Some of the fancier bits included conducting research in the molecular neurobiology lab of a Nobel Prize winner while I was a student at MIT; another involved working for a Silicon Valley billionaire. The not-so-fancy bits included carrying my own poop in a bag to the post office as well as scorching my hair during a botched electronics experiment. Through it all, I kept asking: How do we take the luck out of mental performance upgrades? The best answers I have found, at least so far, lie in the pages ahead.

Back to You

Thanks to improvements in technology and breakthroughs in basic neuroscience, we have probably learned more about the brain in the last few decades than we did in all previous centuries combined. You chose the right time to read a book on brain upgrades.

One of the illusions of modern-day science, however, is that we talk about "the brain" as if it were one thing. In fact, there is no one human brain. Each of our brains is gloriously and maddeningly unique—different from everyone else's. We'll talk about the exciting physical evidence behind this claim in chapter 3, but remember this: if your goal is to upgrade your brain, research on other people's gray matter should intrigue but not convince you. *Just because it worked for them does not mean it will work for you.* Our current understanding of the brain is not sophisticated enough to predict perfectly what will work for each person.

This realization frustrated me to no end. It led to my discovery of something called "self-experimentation." One of the only solutions to the problem of us being so different from each other is to use a scientific method designed *for* the individual. That's not to say we can't learn from each other—you'll find plenty of conventional, group-based research in the pages ahead—but we can't ever assume the transfer will be perfect. To know whether something works for you, you've got to run a self-experiment. This book will teach you to do just that.

You won't need a neuroscience lab to do this. If you can follow a cookbook recipe, you can run a self-experiment. The best part? You'll be taking ownership of the most personal part of you: your brain. When it comes to the health and performance of our brains, many of us feel too shy to share our internal experiences with others. So, who better to upgrade that glorious goo between your ears than you? And if you were ever the kid accused of not working to their "full potential," I wrote this book especially for you.

How to Get the Most Out of This Book

In the pages ahead, you'll read about something called interventions. These are the tools you'll use to change your mental performance. You may be more familiar with this term in the context of trying to help a loved one with a drug or alcohol addiction. In research, the term *intervention* is used more generally. Out of dozens and dozens of candidates, I selected seven interventions to include in this book. All of them are comparatively inexpensive (from free to $500). They can be used at home in 15 minutes a day. Best of all, evidence points to their ability to improve four incredibly important mental abilities: executive function, emotional regulation, learning and memory, and creativity. You'll learn more about those four abilities in Part II.

You'll also find assessments you can use to test your performance across the four mental abilities. These tests are fast to do and you won't need a lot of equipment or a doctor to run them on you. They are also cheap and repeatable, which is important because you'll need to take them multiple times.

I hope this book fills a gap. There are a lot of evidence-based resources for treating medical conditions, but very few evidence-based resources for people without specific diagnoses who simply wish to become better versions of themselves. If you have been diagnosed with a medical condition, you are still absolutely welcome to join me on this neurohacking journey, but please keep working with your doctor, too. Even if you don't have a specific medical condition, I hope you share anything you learn here with your doctor so that they can provide more personalized care for you.

Is It Safe?

Some of the interventions covered here involve gadgets—both hardware and software. Some involve pills. One is mostly a change of mindset. Many were developed by scientists in Western labs, but one was discovered in the South Pacific and others come from Indian and Chinese medicine. For me to include an intervention in this book, there had to have been randomized, controlled scientific research studies on healthy participants that were peer-reviewed and published in academic journals. The participants had to show specific improvements on at least one of our four mental targets. I chose not to include interventions that require a prescription or surgery, or that tend to come with major side effects. The intervention also had to show positive results in humans, not just lab animals.

I've tested almost all of the interventions here on myself. Some have transformed my cognition. Others did less for me personally but proved transformative for other people, so I still included them. Your experiences will be unique. For this reason, it's essential that you learn how to test yourself, which is what we will do in Part I of this book.

IF YOU'RE IN A RUSH . . .

To make it easier to fit this book into the short chunks of time that your schedule may allow, I've provided a time estimate and a stated goal at the start of each chapter. If the chapter takes you less time, congrats! If it takes you longer, that's fine too. Please read at your own pace and in your own way.

At the end of each chapter, there is a set of takeaways. If you only have time to read those, you'll be missing out on a lot of stories and science, but you'll still get the main points.

Once you've read Part I and Part II of the book, feel free to read the intervention chapters (in Part III and Part IV) in any order you want.

The end of the book (Part V) is where you'll find the 15-minute self-experiment protocols. Think of that section as your brain upgrade cookbook. The experiments are organized by mental target, so if you already know which aspect of your mental performance you want to upgrade, you can flip to the relevant target to find instructions for your new 15-minute-a-day routine. The instructions are terse, though; they'll make more sense if you read the chapters, too.

If you ever want to see the source of any statement in this book, there are over 450 citations sprinkled across its pages. Just follow the little superscript numbers you see in the text to the endnotes at the end of the book. The endnotes are organized by chapter. If checking out endnotes doesn't sound that exciting to you, don't worry; you won't miss any of the action if you stick to reading the main text.

By the time you finish the book, you will have:

1. Identified possible bottlenecks holding back your mental performance.
2. Learned how to use personalized tracking and at-home experiments to boost your mental performance.

3. Learned a host of new tools to upgrade your brain. Some will sound like codified common sense; others may seem a bit more...unusual. Like when I attach a battery to my friend's head to help him type faster.
4. Been equipped to upgrade your mental performance using "15-minute self-experiments."

Throughout, you'll learn how my friends and I shared our self-tracking and self-experimentation data with our doctors, in the workplace, and in school settings. Use these stories to inspire you to explore your own cognitive data.

This book is an audacious attempt to throw open the doors to the lab and let everyone come in and play with the equipment that goes beep. Will this cause chaos? Possibly. But good chaos, I hope. You have 24-7 access to something no one else does: your brain. Scientists, doctors, teachers, and technologists can tell you about the averages and characteristics of many brains across crowds, and they have expensive technologies to do so. Still, they don't have full access to *your* brain. If you choose to track and self-experiment, you could become the world's expert on your brain. You could optimize it, tweak it, hone it. In so doing, you could become a profoundly better version of yourself. Who knows what gifts you will give the world then?

Let's get neurohacking!

PART I

Prepare to Neurohack

Chapter 1

Scientific Self-Help

"It doesn't matter how beautiful your theory is, it doesn't matter how smart you are. If it doesn't agree with experiment, it's wrong."
—Richard P. Feynman

> **Time Investment:** 7 minutes
> **Goal:** To understand what scientific self-help is and how to use it to upgrade your mental performance

How much does self-help *really* help?

As a market, self-help is worth somewhere in the billions.[1] Self-help blogs abound, books fly off the physical and virtual shelves, influencers drop their wisdom by the minute, inspirational seminars and workshops attract hopeful hordes...but the impact is rarely measured.

Traditional self-help often involves doing your best to copy what some authority figure tells you to do. The potential flaw in this approach is that this authority figure may be very different from you. Even if they used their techniques to great success, there's no assurance that those techniques will work for you. They may have different personalities or values than you do. They may live in a different environment where their techniques work better.

Another issue with traditional self-help is that it often involves little or no measurement. Measurement would make the whole thing accountable. Often, self-help gurus don't necessarily want you to hold them accountable for whether their ideas actually work.

Scientific self-help is the opposite of traditional self-help. Whereas traditional self-help eschews measurement in favor of feel-good statements and often implies that its way is the best way for everyone, scientific self-help takes a different path. Scientific self-help embraces measurement and accountability—and it assumes individual differences abound. Instead of teaching you how to follow one specific solution, it teaches you a method to compare solutions and decide which one works best for you. Furthermore, scientific self-help empowers you to test solutions for yourself—you won't have to take anyone else's word for it.

The engine powering scientific self-help is the self-experiment. In a self-experiment, the same person acts as the researcher and the subject. For example, if you suspected that your attention would improve immediately after meditation, you could run the following self-experiment: First, measure your attention. Then, meditate for a specified amount of time. Then, as soon as you finish, measure your attention again. Repeat this sufficiently often so that your results aren't skewed by random changes in your attention that have nothing to do with meditation. Sounds simple, right? Doing each of those steps right—and taking steps to avoid bias, since you are playing the role of both the scientist and the subject, after all—is what will transform your idea from a pet theory into a personal discovery.

Trying something, discovering that it doesn't work for you, then trying something else, and discovering that it does—that is the core of scientific self-help. The gloriously imperfect process of science has brought us so much: cures to disease, longer human life, a greater understanding of our physical world, and, in the last few decades, a dramatically expanded knowledge of the matter that makes up our minds: our brains. Just think. If you turn the scientific method back on yourself, what might you discover?

What Is Self-Experimentation?

In research circles, a scientist who gathers data and runs experiments on a single person calls what they are doing "single-case experimental design," "*n* of 1," or "single-subject research." To play the role of both scientist and subject, though, is different. That is self-experimentation. Some call it personal science. Among friends, I call it human guinea pigging.

Do real scientists use self-experimentation? Since the Nobel Prize was first awarded in 1901, we know that at least 14 winners were self-experimenters. Half of them won their prize in the very area in which they conducted self-experiments.[2] Some scientists used self-experiments as a way of proving their confidence in their findings—even at the risk of their own lives and those of their loved ones. Jonas Salk, for instance, actually tested the polio vaccine on himself and his own family—including his wife and kids![3] To some scientists, self-experiments just seemed like the most ethical way to do science. As Rosalyn S. Yalow, the 1977 Nobel Prize winner in medicine or physiology (and the first American woman to win a Nobel Prize in that category) put it, "In our laboratory we always used ourselves because we are the only ones who can give truly informed consent."[4]

Of course, to demonstrate that their findings generalized, self-experimenters had to follow up by experimenting on other people, too. If a professional baker wants to stay in business, they must make bread that other people like. To do that, the baker will offer different types of bread to many people until a recipe is found that the customers crave. But, if that baker needs to please only themselves, all they need to do is bake bread that *they* like. This might sound small or selfish until you realize that if each of us used self-experimentation to unlock the best versions of ourselves, our world would be filled with more competent, more creative, and more compassionate versions of all of us.

Scientific self-help empowers you to be both open-minded and discerning. It empowers you to try out any advice, tip, strategy, or tool and test whether it actually works for you. You may even be able to use scientific self-help to properly test the tips and advice in self-help books, apps, or other tools you already own.

Self-experimentation involves self-tracking. Self-tracking means observing and recording your behaviors without trying to change them. For instance, you might record the number of hours you sleep each night and observe its relation to your mental performance the next day. If I had known about self-tracking as a child, I might have discovered a personal trigger for attention problems: certain types of food. Without proper self-experiments, it took until my 20s—and only after a keen-eyed roommate who suspected I might be gluten intolerant challenged me to focus one of my self-tracking experiments on my diet (more on food triggers in chapter 6, "Debugging Yourself").

Interventions

As mentioned in the Introduction, the term *intervention* is used generally in clinical research. An intervention is like a treatment; it is a tool or approach intended to bring about a specific change. It could be a pill, a meditation program, a new daily ritual...basically, any solution that could be used to address a problem. In our case, we will use it to describe any tool intended to improve mental performance. For instance, it could be a pill aimed at improving memory, a yoga program intended to improve mood, or a video game designed to decrease anxiety.

There's one last term we'll cover for self-experimentation. It's a concept often used by website designers: the "a/b test." That button you just clicked on your favorite website? It was probably a/b tested. The website designers made identical versions of the site, but with small differences: on version a, the button was red. On version b, the button was blue. Then, they waited to see which site got the most clicks. You can run a/b tests on yourself. For instance, you can compare your results on a memory test after 10 minutes of running (intervention a) versus 10 minutes of doing a crossword puzzle (intervention b). While there are a number of ways that running a/b tests on yourself are trickier—you can't clone yourself like the website designers could with their websites, for instance—you can run your experiment enough times and wait a sufficient amount of time between experiments so you can make cautious conclusions about which intervention worked better for you. We'll talk about more ways to make your self-experiments valid in chapter 4, "The Nuts and Bolts."

Now you know what scientific self-help is and how it differs from traditional self-help. You know what self-tracking and a/b tests are. Most importantly, you heard a lot about self-experiments. But how does all of this relate to upgrading your brain?

In the next chapter, you'll get the first piece of that puzzle. You'll learn about a group of people who run a very specific type of self-experiment: neurohackers.

TAKEAWAYS

1. Scientific self-help is different from traditional self-help. It doesn't tell you what to do; rather, it helps you test whether any given self-improvement approach is actually working for you.
2. Scientific self-help involves self-tracking and self-experiments involving interventions and a/b tests. Self-experiments in particular have a long tradition in mainstream science. You'll be using some of the same techniques that Nobel Prize winners have used in their research to upgrade your mental performance.

Chapter 2

Neurohackers, Revealed

"There is only one corner of the universe you can be certain of improving, and that's your own self."
—ALDOUS HUXLEY

Time Investment: 9 minutes
Goal: To see what successful neurohacking looks like

What *Is* Neurohacking?

To some, the word "hacking" is menacing. Computer hackers breaking through online security systems to steal credit card numbers come to mind. To others it means a prank. When I was a student at MIT, our pranks were famous—like the one where students put a police car on the roof of a building, complete with doughnuts in the front seat. Hacking in that context is a way to poke fun, to imagine things done differently. Neurohacking— hacking brain function—involves finding creative shortcuts, using common materials for uncommon purposes, and challenging convention. It is fueled by curiosity—in this case, curiosity about how the mind works.

Neurohacking involves two activities: exploring your current mental abilities and upgrading them. Let's look at an example of how one neurohacker upgraded his memory.

Case Study #1: Upgrading Learning

On September 14, 2010, a computer scientist named Roger Craig became the highest single-day earner on the quiz show *Jeopardy!* ($77,000). He held his title for nearly 10 years. In 2011, Craig explained his approach to retaining the information a *Jeopardy!* champion needs. One of his secrets? A century-old memory technique.[1]

In the 1880s, a German psychologist named Hermann Ebbinghaus shut himself up in a room in Paris to test how memory works. He forced himself to learn, review, and recall nonsense words on a specific, timed schedule. What Ebbinghaus discovered was that the rate of forgetting was predictable. He discovered a pattern of exactly how long it took to forget. If he reminded himself of one of his nonsense words *just* before he knew he was about to forget it—but no sooner—he could save himself hours of studying but still recall the information correctly. The trick was knowing when he was about to forget it. Ebbinghaus's memorization technique became known as spaced repetition. Essentially, it was the most highly specific, scientifically based study schedule you could dream of.[2] Over a hundred years later, specially designed computer programs made following a modified version of Ebbinghaus's schedules feasible.[3]

Our *Jeopardy!* champion, Roger Craig, used just such a program to spectacular success. He found an archive of past *Jeopardy!* questions and answers and input them into the free spaced repetition program Anki.[4] Ebbinghaus would have been proud: after his initial learning period, Craig kept his knowledge fresh and learned new material in just 10–30 minutes a day, focusing only on the information he was *just* about to forget. Using spaced repetition not only saved Craig time, it earned him *Jeopardy!* fame and fortune.

Meet Your Fellow Neurohackers

Whatever goals you have for your brain, neurohacking offers highly personal upgrades to just about anyone. You are about to join the ranks of a bold, inquisitive crew. Prepare to have fun and indulge your inner nerd.

When I started the research for this book over a decade ago, neurohackers were few, and we were sprinkled around the globe. For the most part, we didn't even know about each other. Over time, I interviewed researchers in their labs and built my own tools from spare parts. It took me years to find like-minded individuals who could make that adventure less lonely, not to mention more efficient. You, however, can skip much of that.

Today you can find lively communities both online and in person devoted to biohacking—hacking one's biology (which can, of course, include the brain). There are active groups on Reddit, Meetup, Facebook, and other platforms. Fans of author Tim Ferriss's bestselling book on biohacking, *The 4-Hour Body,* gather online, too.

For a wonderful group of self-experimenters and self-trackers, check out the Quantified Self community. It was started in California by *Wired* magazine founding editors Gary Wolf and Kevin Kelly in 2007. Quantified Self's thousands of members share an interest in "self-knowledge through numbers." Their personal "show-and-tell" projects range from a man who used journaling and self-tracking to lose 200 pounds after a lifetime of obesity[5] to a woman who struggled with fertility and biohacked her way to a healthy pregnancy.[6]

For a community that is focused specifically on the brain and new technologies relating to it, there is the NeuroTechX community, co-founded by a group of North American university students in 2015. Today it hosts "hack nights" (meetups where enthusiasts work on neurotechnology projects together), networking gatherings, and other events for its thousands of members and hundreds of chapters around the world.

These communities vary in how science-oriented their members are, but the range of personalities and backgrounds is broad enough that you'll find someone to your liking if you keep looking. While these groups' membership often reflects the American technology community's demographics—white, male, trained in engineering, fairly affluent—there is a growing portion of people of all races, genders, and professions. What unites everyone? Curiosity. And a belief that understanding your own data can free you of your preconceived notions and biases.

Case study #2: Preventing Brain Freeze

In 2012, Steven Jonas, an analyst at a nonprofit, gave a public talk about his personal project aimed at reducing a stress-related issue with his mental performance.[7] He had noticed that at various points during his workday, his mind froze. Then he would suddenly find himself "fleeing"—skimming news articles and memes, or jumping up for a carb-heavy, pick-me-up snack. Realizing it was a stress response didn't help. He needed a way to physically measure his stress. If he could measure it, he hoped he could learn to manage it.

Jonas found something called HRV, short for "heart rate variability." In a healthy, fit person, the interval between heartbeats varies a lot—high HRV— because it is acutely responsive to changing signals from the brain. Decades of research on the connection between the heart and the brain have produced the finding that, when chronically stressed, the heart becomes less responsive to the brain, leading to low HRV. Jonas decided to track his HRV in order to detect his stress, hoping to catch stress before it turned into a "brain freeze." To measure HRV, you use a chest strap or other sensor worn on the skin. So Jonas modified an old HRV device to beep whenever it detected his HRV decreasing (indicating that stress was increasing).

He began to notice patterns: email set off the beep generally, and emails from certain people really set it off. Soon, he could predict when the beeps would come. This gave him time to try an intervention to reduce the stress: a breathing exercise. With the increased self-awareness that his just-in-time beeps provided, Jonas began to experience fewer brain freezes. As a bonus, his self-tracking revealed that on days when he heeded these just-in-time triggers, he finished his workday with energy left over.

Now's a Great Time to Get into Neurohacking

Self-tracking and self-experimentation—the core of neurohacking—are easier to do now than they have been at any other time in history. We have smartphones with apps that can log your data automatically. We have free spreadsheet tools to document our experiments. You can order many tests and interventions from the comfort of your own home. Even if you prefer pencil and paper for tracking, you can still find online communities in which to get tips and troubleshoot. Doctors are more open to self-tracking than they were a decade ago, when I first began the research for this book. That means that you can (and should!) share your findings with your doctor as you track yourself and run your self-experiments. You can provide data that can help them personalize their care for you.

CASE STUDY #3: CLEARING BRAIN FOG

In the late summer of 2014, Mark Drangsholt, a clinician-scientist and triathlete, gave a talk at a Quantified Self conference.[8] He explained that he had complained to his doctor that he was suffering from brain fog—periods when he couldn't remember words, forgot key information, and couldn't concentrate. Because brain fog can have many causes and because Drangsholt seemed generally healthy, the doctor was unsure how to help.

Drangsholt decided to take matters into his own hands; he gathered genetic, blood, and cognitive test data from consumer companies. Armed with this array of data, he returned to his doctor. Together, they were able to pinpoint the likely cause of his bouts of brain fog: narrowing in small blood vessels in a key area of his brain. The doctor prescribed a statin that lowered his cholesterol levels; Drangsholt's brain fog went away.

Would Drangsholt or his doctor say everyone with brain fog should take a statin? Almost certainly not. Drangsholt's self-tracking, however, helped support a much more personalized form of medicine. Knowledge through self-experimentation is power; Drangsholt's self-knowledge gave him the power to finally dispel his brain fog.

The Neurohacker's Creed

There aren't a lot of rules in neurohacking, but there are four principles of safe and effective neurohacking. I call them "The Neurohacker's Creed":

1. **The neurohacker designs self-experiments.** Neurohackers don't assume that something will improve their mental performance just because someone said it would. They test their own mental performance before and after using it. This testing and evaluation process offers a controllable path toward self-understanding and self-improvement.

2. **The neurohacker picks tests and interventions carefully.** Neurohackers are curious but cautious. They pick the most valid and reliable tests they can, and they test themselves before they try an intervention. If two interventions promise comparable efficacy, neurohackers choose the intervention with the fewest side effects.

3. **The neurohacker never assumes self-experiments will generalize.** Neurohackers know that everyone's brain is different, everyone's life-style is different, and everyone's goals are different. The most successful self-experiments will be highly individualized. Neurohackers can and do learn from each other or from large-scale studies, but they never assume that two people following the same protocol will get exactly the same result.

4. **The neurohacker doesn't have to work alone.** Neurohackers design their self-experiments in concert with teachers, doctors, therapists, and other specialists. Neurohackers take radical ownership of their self-experiments, but they often work with a buddy—a co-adventurer who is also on their own neurohacking path. Neurohacking in pairs or groups keeps everyone accountable and turns self-science into a party!

Now that you've seen examples of neurohackers in action, you may be wondering where to start yourself. Also, you may be wondering what an upgrade really is—and what kind of physical evidence we have that they even occur. In the next chapter, you'll find answers to these questions.

TAKEAWAYS

1. Neurohacking involves two activities: exploring your current mental abilities (that is, through *self-tracking*) and testing different interventions in order to upgrade your mental abilities (that is, through *self-experimentation*).

2. There are numerous online and in-person groups devoted to self-tracking, self-experimentation, and biohacking where new neuro-hackers can find community and ideas.

3. The neurohacker's creed has four principles: (1) the neurohacker designs self-experiments; (2) the neurohacker picks tests and interventions carefully; (3) the neurohacker never assumes their self-experiments will generalize to others or even themselves in the future; (4) the neurohacker doesn't have to work alone.

Chapter 3

The Evidence

"Nothing in life is to be feared, it is only to be understood. Now is the time to understand more, so that we may fear less."

—MARIE CURIE

> **Time Investment:** 11 minutes
> **Goal:** To understand how mental upgrades work and the evidence that they exist, and to start thinking about your own neurohacking goals

In this chapter, we'll look at the evidence that the brain can change, and how. We'll also look at how to measure change given the fact that our brains vary enormously.

Are Brains *Actually* Changeable?

When I first started my research, this question came up often. A version of this is the nature-nurture question: How much of your intelligence is your genetics and how much is from your environment—including the kinds of environments you create for yourself using neurohacking?

Alan Kaufman, inventor of some of the most widely used IQ tests and a clinical psychology professor at Yale University's Child Study Center in the School of Medicine since 1997, has said, "Probably the notion of genetics contributing about 50 percent to IQ and environment contributing about 50 percent to IQ is as close as we are going to get within science to estimating the relative role of each in a person's IQ."[1] To consider the effect of environment on the brain, Kaufman and colleagues compared the IQs of siblings who live together to those of siblings to who live apart. They found that siblings who were reared together but who were not biologically related to each other (that is, adopted siblings raised together) had fairly correlated IQs (a correlation of .28). As adults living apart, their IQs diverged: after they no longer shared an environment, their IQs' correlation was a measly .04.[2]

These findings have limitations, of course. For one, these were observational studies, not lab experiments. In observational studies, scientists just record what happened, and notice that certain things tend to happen after other things happened—and if that kind of thing happens often enough, probably those two things are related and it wasn't just chance. Only when you run a proper experiment can you make strong statements about cause and effect. In this case, of course, it would be unethical to run such an experiment. For another, IQ is a limited and highly imperfect proxy for mental performance. All of these criticisms aside, the correlation studies about IQ are worth at least some cautious optimism. They show that IQ is not just a puppet to genetics. It can and does change in response to its environment. Given the right environment, it stands to reason that it should be improvable. In other words, this is evidence in favor of neurohacking. But how exactly does the environment enact changes on the brain?

There are numerous mechanisms proposed for how the environment changes your body—including your brain. For instance, changes in the environment can affect how your genes are expressed (as studied in the field of epigenetics). Another way is through the microbiome—the ecosystem of microorganisms living in and on your body, which can affect everything from mood to energy levels. This can be altered by, for example, food and stress. The last—and most frequently mentioned way—is that our brains can physically change in response to experience.

Neuroplasticity

We won't get too technical here, but it's worth understanding the basic evidence underpinning the belief that we can change our own brains. Neuroplasticity is defined as the brain's ability to form and reorganize connections in response to learning, to new experiences, or to injuries. Our brains change physically in response to negative things (like childhood adversity or stress) and to positive things (like learning opportunities).

As neurohackers, we harness the brain's ability to change through learning. As we learn, our brains change in predictable ways. For example, in 2014, a Korean research team recruited three groups for a brain imaging study: college students with no archery experience, collegiate archers, and Olympic medalists in archery.[3] When novices "mentally released the bowstring," imaging showed widespread activity across their brains. In contrast, the collegiate archers made far more efficient use of their brain, engaging fewer regions; the Olympic archers' brains engaged the fewest regions of all.

The connections between neurons are referred to as wiring. Neuroscientists use the phrases "neurons that fire together, wire together" and "neurons that fall out of sync lose their link" to explain why some neurons get connected to other neurons and why they disconnect. When you learn a new scale on the piano or remember an equation for the first time, you are literally making a set of new neuronal connections. The formation of new connections (wiring) between neurons is known as synaptogenesis.

When trying to learn new tasks, neighboring neurons often get recruited to help. If neurons near the ones principally involved in processing a learning task aren't too busy with other responsibilities, they get roped into helping out with the learning task, too. As humans develop expertise in an area, not only does the communication between large-scale brain regions change, so does the real estate used to accomplish the learning task. Learning a musical instrument,[4] juggling,[5] memorizing the drivable layout of a city,[6] cramming for standardized tests in law[7] and medicine,[8] and undergoing talk therapy[9] have all been found to cause observable, physical changes in the brain. All this is to say that brain change is very possible.

Our Brains Rewire Themselves All the Time

Not only do we change in response to learning new tasks, we change in response to our environment and our life experiences more broadly. A professor from my department at MIT, Sebastian Seung, introduced me to the power of the "connectome"—the precise and unique wiring of different neurons connecting to each other in the brain. In a multi-hundred-person study, American and Taiwanese researchers found, on average, that each person's brain was more than 12 percent different than it had been 100 days earlier.[10] The brain's ability to rewire itself is, at least in part, what allows us to learn and change, even as adults. All the more reason to make your own brain upgrades intentional; then you get to ensure that the rewiring is happening in ways that you want.

How Do We Measure Brain Change?

To answer this, you'll need to understand how mental performance is measured. There are two ways: behaviorally and biologically.

Testing mental performance: Behavior tests

The behavioral tests I recommend using for your self-experiments are those that measure mental ability (not cultural knowledge) and can be repeated. Since our goal as neurohackers is to use tests to gauge our progress, we need tests that are valid even when repeated, so we can see differences in our performance over time. If every time you take a test you get more familiar with its questions, a higher score later doesn't mean the same as a higher score earlier. We want to avoid that. Useful tests often involve visual puzzles or games. In this book, I will give you shortened versions of some of these tests. I've selected ones that you can take repeatedly so you can assess the effects of your self-experiments. We'll talk about them more in Part II of this book.

There are tests I recommend *against* using. Many commonly used personality and intelligence tests, for example, are neither repeatable nor culturally unbiased. Thus, you can end up with a low score due to simply not knowing culturally specific vocabulary. For instance, here's an old question from the SAT—a test that correlates highly with IQ tests and that used to be very culturally biased.

RUNNER: MARATHON
(A) envoy: embassy
(B) martyr: massacre
(C) oarsman: regatta
(D) horse: stable

In this case, the correct answer is C. (An oarsman is an athlete who rows a boat, and a regatta is a boat race.) These words might have been familiar to a certain upper-crust set, but not to very many outside that set. That question—and others like it—were later deleted from the test for being culturally biased.[11] This is the kind of test we want to avoid.

TESTING MENTAL PERFORMANCE: BIOLOGICAL TESTS

Measuring physically observable changes in the brain—such as changes in the brain's electrical activity—is probably the future of neurohacking. Cheap wearable brain imaging devices offer an exciting glimpse at where the field will go, giving you a chance to observe your brain's functioning in real time. Brain imaging has also been used to diagnose ADHD; one FDA-approved application allows doctors to compare a pediatric patient's brain waves to the brain waves of children definitively diagnosed with ADHD.[12] However, most of the medical-grade biological mental performance assessments are either not yet advanced enough or not accessible enough to enable neurohacking at home. That mostly leaves us with the behavioral tests in the meantime.

Individuality, Personalization, and Choosing Your Neurohacking Goals Wisely

At this point you may be wondering: What does the end goal of neurohacking really look like—that is, what does an upgraded brain look like? The answer is that it will look different for each person. As a neurohacker, you're in charge of your own upgrades. You get to make your own self-experiments, and you get to decide what counts as optimizing your mental performance. Just because you can change something about yourself doesn't automatically mean that you should.

Neurodiversity is a buzzword you may be familiar with: it refers to the naturally occurring diversity of all our gloriously different individual human brains. Just as there is a gorgeous variety of human hair, skin, and eye colors, there is even greater variety in the brains behind those faces. Individual differences in mental performance can be significant: no two brains are identical, not even those of identical twins.

Recent imaging data has shown us that the wiring pattern of your brain is unique and identifiable,[13] even over time—similar to fingerprints.[14] Unlike their genetics, the brain wiring of identical twins is not identical. Researchers in the US and Taiwan found that identical twins' brain wiring was only about 13 percent similar. For context, fraternal twins and siblings wiring overlapped by about 5 percent.[15] A brain's electrical patterns may be unique, too: when a group of study participants were faced with the same 500 images (including a slice of pizza, a boat, a picture of actor Anne Hathaway, and the word *conundrum*), each of their brains' responses was so different that the researchers were able to identify each participant with perfect accuracy.[16]

Neurodiversity is an idea that has sparked a social movement as well as a scientific movement. People whose thinking and behavior have been labeled "abnormal" have started to organize and advocate for acceptance of their mental traits. One of the largest such groups is composed of people diagnosed with autism spectrum disorder, but people labeled with ADHD, bipolar disorder, dyslexia, and many others are also advocating for acceptance rather than pathologizing of their differences.[17]

They argue that their brains are wired differently, but not always worse; under certain circumstances, they have advantages over so-called neurotypical people (those whom others might label "normal" or "healthy"). They also argue that some of the differences currently labeled as disorders may be explainable as variations that evolved to help humans adapt to different circumstances, just as, for example, different skin colors became prevalent in environments with different amounts of sunlight (dark skin provided humans with better protection against direct sunlight in areas close to the equator, but pale skin enabled better vitamin D absorption in northern areas).[18] For instance, some people with ADHD hypothesize that they are descended from highly successful hunters (rather than gatherers). While many people with ADHD struggle when they have to juggle too many low-intensity (boring) details, their ability to hyperfocus in situations that others find overwhelming may have helped them when closing in on prey. This may explain, for instance, the relatively high number of people with ADHD who end up as ER physicians.[19] Similarly, the ancestors of "night owls" or evening-type chronotypes—those whose natural inclination is to fall asleep and wake up later than people with morning chronotypes—may have been on the night watch, essential for alerting the rest of the tribe to threats. Interestingly, this group of people has scored higher in lab tests on multiple measures of mental performance when tested at their best time of day (evening) — higher even than other chronotypes when tested at *their* best times.[20]

So, as you go about your neurohacking journey, you may discover things about your mental performance that seem, at first, to be disadvantages. Before jumping to "correct" them, I urge you to consider what positive role they may be playing — possibly even secretly! It's your mind and you should feel free to upgrade it as you wish, but personally, I hope you do *not* use neurohacking as a way to make yourself indiscriminately more similar to the people around you. This world has a lot of problems and opportunities, and your brain may very well be uniquely equipped to address one or more of them. I urge you to use neurohacking to make your brain a better version of itself — but still very much itself!

To understand *how* to actually upgrade your brain, you'll need to learn how to run self-experiments. The next chapter has you covered.

TAKEAWAYS

1. Brain imaging evidence has shown that the human brain can be upgraded intentionally, even in adulthood. A person's brain wiring changes significantly over time. Experts pin intelligence at roughly 50 percent environmental, 50 percent genetic. This is good news for us since neurohacking is about making intentional changes to your environment.

2. Individual differences are enduring characteristics that differ across people, such as their fingerprints and their brain patterns. Individual differences matter when it comes to the brain. They are one of the reasons that one-size-fits-all neurohacking interventions are unlikely to work; personalization is key.

3. Neurodiversity is part of human diversity. You can use neurohacking to make yourself more neurotypical or you can use it to strengthen what makes you unique. You could do both, depending on the time frame and on the particular self-experiment. It's your brain, so it's your upgrade.

Chapter 4

The Nuts and Bolts

"When you can measure what you are speaking about, and express it
in numbers, you know something about it."
—Lord Kelvin

> **Time Investment:** 12 minutes
> **Goal:** To learn the steps in a neurohacking
> self-experiment

I've messed up many neurohacking self-experiments. One time, I forgot to measure my mental performance before I started—it wasn't until I reached the end of the experiment that I realized I had no baseline to compare against! Another time, I ran a self-experiment for so few sessions that I couldn't tell whether my results were due to chance. Frankly, my blunders could fill another whole chapter. It won't be this one, though. My goal here is to save you time by avoiding those same mistakes—so, let's begin!

You're about to get a framework for neurohacking self-experiments. You'll even get a sample schedule at the end of the chapter. If you follow it, you'll be far more likely to pick a good mental target to upgrade, choose a good design for your self-experiment, and ultimately, upgrade your mental performance more quickly.

Now, it may sound odd, but I want you to imagine a ladder...

The Four Steps of the Neurohacker's Ladder

Imagine a ladder with four steps leaning against a wall. On the other side of the wall is an upgraded version of your brain. The trouble is, you're on *this* side of the wall. You need to climb the ladder to get to the other side. You notice that each step has a word painted on it:

- Focus
- Selection
- Training
- Reflection

Taken together, Focus-Selection-Training-Reflection can be abbreviated as F-S-T-R. I pronounce them "faster." I think of the four steps of the neurohacker's ladder as encompassing the four key phases of a neurohacking self-experiment.

Before you start climbing, you'll want to prep:

1. **Get a lab notebook.** In chapter 5, you'll learn how to pick one.
2. **Find a neurohacking buddy if possible.** In chapter 5, you'll learn who to pick and how to work together.
3. **"Debug" yourself.** In chapter 6, you'll find potential bottlenecks in your health and lifestyle that could be holding back your mental performance.

Now, let's climb that neurohacker's ladder. We'll start with the first rung, Focus.

F is for Focus

The F in F-S-T-R is for "Focusing (your goals)." If you try to improve too many things at once, you'll end up spending far more than 15 minutes a day. Spend those minutes wisely, otherwise you won't run a very scientific self-experiment. Many pitfalls await the careless neurohacker. They include such gotchas as bias, practice effects, and carryover effects. Any of these technicalities can snag your ability to figure out whether an intervention truly worked for you. Here are steps to mitigate those risks:

1. **Pick a mental target.** To do this, you'll want to assess which mental abilities are currently your strongest and which are your weakest. Part II of this book describes how to assess yourself across four domains of

mental performance: executive function, emotional regulation, learning and memory, and creativity. Each has strong relevance to mental performance in daily life. There's one problem: evaluating yourself—rather than using a performance-based measure—carries a strong risk of bias. Whether due to ego or wishful thinking, it can be easy to grade your mental performance in rosier terms than you probably deserve. Conversely, if you're feeling depressed or are a particularly harsh self-critic, you might grade yourself unfairly low. There are a few ways around this problem, thankfully. One is that you'll be using a variety of measures to assess yourself, some of which are less subjective and more performance-based. So, your likelihood of skewing everything with your own bias is at least somewhat managed.

2. **Gather data on your baseline mental performance and your baseline quality of life.** To have some basis of comparison, you will need to establish a baseline for your mental performance *before* doing any interventions. Remember our "gotchas" list of neurohacker pitfalls? Let's deal with the risk of so-called practice effects. Unlike stepping on a weight scale where you can't get a dramatically better number simply by testing multiple times, we tend to get better at cognitive tests the more times we take them. This is because we get more comfortable with the format, not because our underlying cognition has actually improved. To deal with practice effects, you have a few choices. One option is to take a cognitive test at the beginning of your experimental program and then wait long enough before taking a follow-up test that you've mostly forgotten the questions; many clinicians advocate for a six-month interval between tests.[1] I recommend a different approach, one I first saw used by Yoni Donner, the Google researcher behind the online cognitive testing platform Quantified Mind.[2] In this approach, you test yourself enough times that your answers stabilize. Complete the Focus phase within a week, taking each of the performance-based tests about five times.[3] Save the second-highest score you achieve during the baseline period, since your highest score may have been a fluke. This second-highest score will be your baseline score to beat during your self-experiment period.

Another way to make sure that your interventions are actually working for you is to look for changes in your day-to-day life. You'll assess your Life

Satisfaction score and your Say to Do score in chapter 12. These two scores are intended to give a rough baseline of your current quality of life and productivity levels. If you compare your scores in these areas before and after your self-experiment, you'll have "real world" metrics with which to assess your neurohacking efforts. Just because one thing happens before another thing doesn't mean that one caused the other (that is, correlation doesn't equal causation). However, if your Life Satisfaction and Say to Do metrics go up along with your mental performance improvement, it's at least possible that your neurohacking helped improve all of them. Again, you'll read more about this in chapter 12, "Life Scoring."

S IS FOR SELECTION

The S in F-S-T-R is for "Selecting (an intervention)." During this phase, you will select a self-experiment, do your prep, set up your experiment, and order your tools. Here are the actions you'll take as part of this stage:

1. **Choose your intervention.** Parts III and IV introduce you to different interventions you can use to improve your mental targets. Most studies assume you will test one intervention at a time. You could combine interventions, however. For instance, a large-scale, randomized controlled study of a few thousand older adults in Finland resulted in impressive cognitive gains by combining diet, exercise, computer-based cognitive training, and health monitoring.[4] Note that if you choose a "kitchen sink" approach, as my friend and Stanford professor Irina Skylar-Scott likes to call it, you won't know which of the interventions was most responsible for your improvements.[5] Still, you *can* use a self-experiment to know whether that particular kitchen sink helped or hurt your cognition.
2. **Choose or design your particular self-experiment protocol.** Most of this work is done for you, since Part V lists self-experiment protocols organized by mental target. There is also information on the cost, complexity, and materials required, as well as instructions on how exactly to run each self-experiment. All of the protocols are designed to take roughly 15 minutes a day and to range in cost from free to under $500.

Let's return to our "gotchas" list of neurohacker pitfalls. Our last risk is the so-called carryover effect. This occurs when there are lingering effects of a previous experiment on a current experiment, making it unclear which intervention deserves the credit for an improvement. You can handle this in a few ways. One way is to leave long gaps between interventions. This is called a "washout" period.[6] By not using any interventions for a while, you "wash out" the effects of one intervention so that your system is "clean" before you try another intervention. The sample self-experiment calendar at the end of this chapter includes a washout period.

Most of the self-experiments in this book, however, are designed to help you compare the *immediate* effects of one intervention to those of another, rather than attempting to estimate the longer-term effects. Even if you had lingering effects from an intervention you did yesterday, the effects of an intervention you *just* finished are likely to be far stronger than those lingering effects. Most of these self-experiments answer the question "If I had just 15 minutes to give myself a quick mental pick-me-up, which intervention would work best for me?" They compare your test results after the intervention to those right before.

If you're not interested in immediate effects, however, you need a different approach. For instance, one self-experiment evaluates whether a particular herbal supplement helps improve your memory. This particular herb, however, is known to take multiple weeks to kick in (see chapter 19, "A Pill a Day"). In this case, taking a daily test before and after consuming the herb would make no sense. Instead, the self-experiment calls for testing yourself before you start taking the herb and again a few weeks later, after the pill is expected to have taken effect. Furthermore, you can't alternate with any other intervention for an experiment like this, because the carryover effects would be too hard to decipher.

1. **Choose a randomization schedule.** The simplest option here is to simply alternate which of two interventions you do each day. The downside of the alternation approach is that you may introduce what's known as systematic bias. For instance, say you always ran intervention A (say, 10 minutes of exercise) on Mondays and Mondays are stressful for you. Furthermore, let's say you always ran intervention B (say, 10 minutes of meditation) on Saturdays—a day that you typically

do something fun. For reasons that had little to do with the actual efficacy of the interventions, you would likely end up finding that your mental performance was worse after intervention A (exercise) than intervention B (meditation). So, to avoid systematic bias, you can use a method that statisticians call "sampling without replacement." In our example, you would use a special method of picking whether to exercise or meditate each day. The method ensures that you never know which intervention you're going to use before you use it. You also end up using both interventions an equal number of times. This maintains suspense, which can be fun, but it does have downsides, which we'll talk about in Part V.

2. **Choose an experimental program length.** To avoid getting fooled by randomness, you'll need to run your interventions multiple times. Exactly how many times depends on how strong the intervention's effects are and how easy it is to mistake chance effects for real changes. In general, I recommend testing each of the interventions in this book at least 15 to 30 times. If you did daily intervention sessions, you could run multiple self-experiments from the beginning to the end of each quarter of the year. You'll learn more about this in Part V.

3. **Buy or build your tools.** To prepare for your exercise self-experiment, check the materials list to see what you need to buy or build.

T IS FOR TRAINING

The T in F-S-T-R is for "Train (your brain)." This is when you actually use the interventions and follow the protocols. Here are the actions you'll take as part of this stage:

1. **Follow the protocols for the length of time you've chosen.** Use the randomization method you chose. Note your pre- and post-test results in your lab notebook each day.

2. **Allow for a "washout period," while still collecting data.** As mentioned, this helps eliminate the problem of carryover effects. However, you still need to take all of the mental performance tests available for your mental target each day of the washout period to see how well you

do without the intervention, just as you did during the baseline period. The purpose is to see whether you improved not just on the test you used during the training period, but also on the other tests of that mental ability that you practiced less. You'll also reassess your life scores. Finally, you'll take all four of the self-assessment surveys to see if your mental performance in other domains has changed since the beginning of the self-experiment.

R IS FOR REFLECT

The R in F-S-T-R is for "Reflect." This is when you look back at the data you gathered in the baseline, intervention, and washout periods. Here are the actions you'll take as part of this stage:

1. **Graph your data.** When you want a quick, general sense of whether a particular intervention is working for you, there's a lot you can learn from plotting your data graphically. When you see the data graphed, it helps you decide which intervention is working better, so you can decide whether to keep using both, scrap one, or try something fresh for your next experiment. Analyzing your graphs visually rather than as calculations is my recommended approach. You'll learn more about this in Part V.

2. **Interpret your data.** If you prefer to make a judgment based on numbers rather than graphs, you could start with a very simple approach. Simply pick the second-highest score you earned in each of your experimental periods—your baseline, intervention, and washout periods—and compare those three scores to each other. (You take the second highest, because the highest score is likely to have been a fluke.) A more conventional but more calculation-intensive way would be to compute some statistics. You could calculate the average and standard deviation and pick a confidence interval for your baseline scores. Then, do the same for the intervention period test scores and for the washout period test scores. Finally, compare the three means (with their confidence intervals) to each other. In self-experiments, it's easy to get fooled by these kinds of numbers, however. You'll read more in Part V about why pretty pictures are often better than statistics.

3. **Decide what to do next.** Based on your data, you'll consider whether to stay focused on the same goal or pick a new one. You'll decide whether to use the same or a different set of interventions for your next self-experiment.

A Sample Calendar for a Self-Experiment

Here's a sample schedule for conducting a neurohacking self-experiment. It shows an example period during which you would self-track and try the

Sample Program	Sunday	Monday	Tuesday	
Week 1: Baseline Period—Focus	Pick a mental target	Debug yourself	Baseline	
	–Take all four self-assessments of mental performance (surveys) –Get a lab notebook	–Take Health & Lifestyle (HL) survey	–Take mental performance tests in target area (optional: HL tracking)	
	☐	☐	☐	
Week 2: Prep Period—Select	–Choose protocol focused on mental target (optional: Say to Do tracking)	–Choose protocol focused on mental target (optional: Say to Do tracking)	–Buy/build tools (optional: 1. Say to Do tracking; 2. Find a neurohacking buddy)	
	☐	☐	☐	
Weeks 3–10: Intervention Period—Train	Exercise	Rest	Exercise	
	–Take pre-test, do exercise, take post-test –Say to Do tracking	–Take pre-test, do rest, take post-test –Say to Do tracking	–Take pre-test, do exercise, take post-test –Say to Do tracking	
	☐	☐	☐	
Week 11: Washout Period	–Take mental performance tests -Assess Life Satisfaction score	–Take mental performance tests -Assess Say to Do score	–Take mental performance tests –Take Health & Lifestyle survey	
	☐	☐	☐	
Week 12: Reflection Period—Reflect	Graph intervention data	Graph washout data	Compare intervention, washout, and baseline graphs	
	☐	☐	☐	

interventions. The length is a trade-off between my impatience (wanting you to get results as soon as possible) and my desire for accuracy (wanting you to gather data for long enough that you've reduced the likelihood that you're just seeing an effect due to chance). Some experiments can be done faster than the example I've given below, and some will take longer. The calendar should give you a sense of each of the phases, though: baseline, prep, intervention, washout, reflection. Note that you'll see some terms you may not recognize yet (like "debug yourself"), but you'll learn all about this—and get access to the surveys mentioned below—in chapters 6 through 12.

Wednesday	Thursday	Friday	Saturday
Baseline & Life Satisfaction	Baseline & Say to Do	Baseline	Baseline
–Take mental performance tests in target area (optional: HL tracking) –Assess Life Satisfaction score	–Take mental performance tests in target area (optional: HL tracking) –Assess Say to Do score	–Take mental performance tests in target area (optional: HL tracking, Say to Do tracking)	–Take mental performance tests in target area (optional: HL tracking, Say to Do tracking)
☐	☐	☐	☐
–Buy/build tools (optional: Say to Do tracking)	–Buy/build tools (optional: Say to Do tracking)	–Buy/build tools (optional: Say to Do tracking)	–Buy/build tools (optional: Say to Do tracking)
☐	☐	☐	☐
Rest	Exercise	Rest	Exercise
–Take pre-test, do rest, take post-test –Say to Do tracking	–Take pre-test, do exercise, take post-test –Say to Do tracking	–Take pre-test, do rest, take post-test –Say to Do tracking	–Take pre-test, do exercise, take post-test –Say to Do tracking
☐	☐	☐	☐
–Take mental performance tests	–Take mental performance tests	–Take mental performance tests	–Take all four self-assessments of mental performance (surveys)
☐	☐	☐	☐
Choose new focus area	Choose intervention	Design intervention protocol	Buy/build tools
☐	☐	☐	☐

Phew! You just made it through one of the tougher but most important chapters in this book. You now know what's involved in running self-experiments. You learned how to climb the Focus-Selection-Training-Reflection (FSTR) ladder to a world where there's an upgraded version of your brain. I hope the example template above will help guide your path, but remember, this is just one example. Again, not every experiment will take 12 weeks—some can be done faster, and some take longer. Also, while you should certainly try to be careful, you won't be able to control everything. Certainly, if you just ran a marathon, it might not be the best day to try to assess how an additional 10 minutes of exercise affects your mental performance. Use your judgment, record your observations as you go, and enjoy yourself. As in any new adventure, there will be parts where you feel uncertain or you lose drive. In the next chapter, we'll discuss proven strategies you can use to keep up your motivation and stay organized even when the going gets grueling.

TAKEAWAYS

1. The four steps in the Neurohacker's Ladder are: Focus-Selection-Training-Reflection (F-S-T-R).
2. Certain types of technical "gotchas" reduce the validity of self-experiments. Those include risks like different types of bias, practice effects, and carryover effects.
3. There are specific recommendations to address these issues, such as how to test your mental performance, how long to wait between experiments, and how long to run an entire self-experiment.
4. The sample self-experiment calendar in this chapter provides a timetable within which to follow the recommendations.

Chapter 5

Organize to Motivate

"By failing to prepare, you are preparing to fail."
—Benjamin Franklin

Time Investment: 11 minutes
Goal: Learn how to stay motivated and organized
during your neurohacking experiments

Since 2011, I've been tracking my New Year's resolutions, as well as my general satisfaction with my life. To do this, I carve out some time alone during the first week of January, and I go through a thorough review of the previous year. To date, it is one of my longest running self-tracking projects. Over its course, I've gone through job and relationship changes, moved across the country, gone to grad school, written this book, and had a baby. When I reflect on why this self-tracking project had such staying power when others fizzled out, I realize I used three approaches to make it a habit: (1) I use a special notebook for it, (2) I have an accountability buddy, and (3) I use evidence-based motivation hacks to keep myself going.

Using these three tools has helped me remember to self-track, use my interventions when I'm supposed to, and review my data often enough that I incorporate lessons into my life going forward. In this chapter, we'll explore these three tools, as well as a few other research-backed tidbits on motivation and organization. I hope they make neurohacking a habit for you, too.

Use a Neurohacker's Notebook

Monitoring is medicine. Just recording and measuring can make you more aware and improve in and of itself. To do this, you'll need a designated notebook. You'll use it to record your goals (such as your mental targets), your decisions (such as which interventions you'll be testing), and your daily activities. For instance, you'll record when you test yourself, when you use interventions, what day and time you use them, and any related observations. You will make graphs in it, and you may do some calculations there, too.

What kind of notebook works best? I like to use online spreadsheet programs like Google Sheets so I can access my data anywhere. Others prefer desktop spreadsheet programs for privacy reasons. Some prefer to go more old-school with a physical lab notebook. If you do this, be prepared to do any relevant graphing or calculations by hand. If you feel more inspired by blank paper rather than cross-hatched graph paper, go for it, but be advised that you'll probably end up printing out graphs and pasting them into the book.

Since this is personal information, I recommend storing any spreadsheets and documents related to your neurohacking experiments in a password-protected folder on your laptop or in the cloud. A firewall, frequent virus scans, and backups are good ideas, too.

Find an Accountability Buddy

Throughout the book, I will refer to "accountability buddies," but feel free to use an "accountability chain" instead. This is where you report to one person and they report to a third person. If that third person reports to a fourth person, and the fourth to a fifth...the accountability chain could continue on and on!

The why. While it is absolutely possible to run the self-experiments in this book on your own, it is likely to be easier, more fun, and more effective if you recruit an accountability buddy. Having an accountability buddy who you check in with and send progress reports to could nearly double your goal achievement. In 2015, psychologists at Dominican University of California wanted to know how the act of writing down goals might change people's goal attainment over a four-week period.[1] They found that participants with an accountability buddy and written goals succeeded in reaching their goals significantly more often than those with only one of the two or neither. Those who told a friend about their goals and sent that friend weekly progress reports succeeded at the largest percentage of their goals.

You'll notice that some of the following tips on finding and keeping an accountability buddy are basic and slightly opinionated. I promise I'm not trying to send you back to kindergarten. If you failed sharing and couldn't stop hitting people in the block area, that's your business. But following these suggestions will not only improve your odds of having an enjoyable partnership, they're likely to increase your odds of successful neurohacking experiments, too.

The how. Once you've found the right accountability buddy (more on that in a moment), try these tips for how to work together productively:

> **Share *relative* data, not *raw* data.** While I love sharing the neurohacking journey, I've never felt comfortable sharing my raw data with my buddy. If you feel similarly, here's a trick. When I share mental performance data with my accountability buddy, I share only what I call the "percent change." That is, I share only the amount that my scores have changed each week relative to my baseline scores. My buddy does not know either my baseline scores or my current scores. She sees only numbers like "15 percent increase" and "10 percent decrease." My buddy never knows my exact score on any particular test, but she can provide feedback if she sees my scores trending in a particular direction.

> **Make check-ins a habit.** It's natural for good intentions to slide as the high of New Year's resolutions wears off. This is why my accountability buddy and I hold check-ins at the same time and day each week. At our meetings, each of us does three things: report on how our previous week's neurohacking went, report on what we plan to do the following week, and then pose a question to discuss. I might ask my buddy, "Can you text to remind me to do my intervention before I go to bed?" or "I'm torn between testing out a video game or trying neurofeedback; would you be willing to try one of them with me?" If I'm the one to cancel a meeting, I try to be the one to reschedule it, too.

> **Give as much as you get — gently.** Finding a balance between encouragement and useful feedback is key. My accountability partner and I rely on each other to provide gentle reminders of our previous resolutions. I'm

reminded of an old saying: "A friend is someone who knows the song in your heart and can sing it back to you when you have forgotten the words." Done the right way, feedback feels refreshing, not like nagging or judgment. We've found it helpful to have explicit conversations about communication so that each person can receive feedback in the way that works best for them. Personally, I like the "shit sandwich" approach to feedback: the bottom piece of bread is some kind of true and positive observation of what I am doing well, the shit (which is the meat in the sandwich) is some constructive suggestion (that is, something I could do differently, ideally with a concrete suggestion), and the top piece of bread in our metaphorical sandwich is another piece of positive feedback or a genuine affirmation. Feel free to experiment if you're not sure what communication style works best for you.

The who. I'm a huge fan of having an accountability and/or general neurohacking buddy. However, it's important that expectations are clear and shared between the partners. My first few accountability partners didn't work out well. As I discovered, great romantic chemistry with a significant other, an old friend who makes you laugh until you cry, and family members who you'd give your kidney to don't necessarily equate to good accountability partners. A few years ago, I was lucky to make a friend who also became a wonderful accountability partner. In contrast to my previous attempts, this partnership felt simple: we shared a similar level of motivation to show up to each meeting prepared and a willingness to be open and take responsibility for both our successes—like hitting writing or business goals—and failures—like not cleaning the house as promised.

For a maximally successful buddyship, I have found it best to be compatible in four domains:

Motivation: Pick an accountability buddy with similar levels of motivation and ambition. If one person is more casual in their approach and the other is devoting their life to the endeavor, friction will definitely occur! The key is to be aligned on your level of commitment. Do both of you want to spend similar amounts of time each week on neurohacking? Are both of you hoping for a similar amount of change in yourselves? Note that you don't need to have the same goals for these to be true.

Work style: You'll gel best if you make sense of information in compatible ways. If you're hyperorganized and love lists and schedules, you may get annoyed by a partner who likes to keep everything in their head. Someone who prefers spontaneous phone calls over planned meetings will be irritated by a schedule fanatic. Lovers of physical notebooks will feel frustrated by those who adore all things digital. Find someone who likes to work in the ways you work, and you will spend less time discussing *process* and more time making real *progress*.

Experience: If you're already experienced and confident from previous self-improvement projects — say, you've successfully lost weight or learned a hobby, a musical instrument, or a foreign language on your own — you may want to pick a buddy who has a similar experience level. If you're new to all of this, you may want to buddy up with someone who is also a beginner so you can learn together, or you may want to find a more experienced buddy who is excited about mentoring you.

Trust and respect: Pick an accountability partner whom you trust and care about. Trust is crucial because you'll be sharing your mental performance with them — neurohacking is a very personal thing (although, as I mentioned earlier, sharing *relative* data rather than *raw* data definitely helps mitigate this). Finding someone who will neither feel jealous if your performance is better than theirs nor make you feel bad if your project isn't coming along as quickly as theirs can be tricky. If you fundamentally wish them well — and they feel the same — you'll both feel inspired to do your best. Your buddy doesn't have to be your all-time hero, though — in fact, it may be better if they're not. After all, you want to feel free to try new things, share mistakes, and learn with them.

Making Habits Stick

Here are some methods to help you stick to the new practices you're trying to start.

Make an implementation plan: One of the most powerful predictors of a goal actually being achieved is whether there is an "implementation plan." That means a plan for where, when, and how you will pursue your goal, as well as how you'll respond to possible distractions and obstacles. A Harvard research team found that implementation plans helped students who were at risk of dropping out make it to school more reliably. They're also now a common approach for voter engagement teams to improve turnout for elections.[2] Simply asking yourself "What is my plan?" turns out to be surprisingly powerful.

To apply an implementation plan to your neurohacking project, answer the following questions:

1. *Where* will I do my neurohacking each day?
2. *When* will I do my neurohacking each day—and for how long?
3. *What* materials will I need to do my neurohacking each day?
4. *What* kinds of obstacles or distractions do I predict will prevent me from doing my neurohacking each day? For each of those obstacles or distractions, what can I do to prevent them from derailing me?

One of the reasons I've stuck with my long-running self-tracking project is that I keep to a yearly schedule: my resolutions happen every January, I set milestones at the beginning of each quarter, and I set weekly and daily goals, too. As you begin your self-experiments, make sure you know what you'll be doing on any given day. Refer to the sample schedule you got in chapter 4, "The Nuts and Bolts." It will help make your neurohacking plan clearer.

Stack habits: One of the keys to success when starting a new habit is to piggyback or "habit stack" on an existing, pre-established habit. This is what researchers call a "keystone habit." For example, if you want to build the habit of neurohacking for 15 minutes a day, and you already have a very strong habit of eating breakfast every morning, you should plan to do your neurohacking directly after breakfast. The keystone habit will provide the trigger for you, so you won't have to remember to do it, you'll just do it automatically after the other thing you do automatically.

Treat yourself with flexibility and compassion: Sometimes we question whether we really have what it takes to accomplish our goals. What's the antidote? Self-compassion. In studies of the most resilient people, the ones who do best are those who accept temporary setbacks without beating themselves up about it. As one of my favorite lines from the 2005 movie *Batman Begins* goes, "Why do we fall, sir? So that we can learn to pick ourselves up."[3] Resilience isn't never failing; it's learning from failure until you finally succeed.

To explore this idea, a group of researchers tested three incentives for groups tasked with completing boring tasks each day.[4] One group got rewarded even if they completed only 5/7 of the days (the "easy" group), one group only got a reward if they completed all 7 days perfectly (the "no excuses" group), and one got the "no excuses" rules, but with a second chance (if they needed it, they could have two extra days to complete their tasks). If you guessed that the "second chances" group performed the best, you were right. The "no excuses" group performed worst; only 21 percent of its members earned the reward. The "easy" group came out just slightly better, with 26 percent earning the reward. Of the "second chance" group, however, a whopping 53 percent earned the reward. Moral of the story? Aim high, but give yourself a second chance if you mess up—you're more likely to stick with it. Researchers have also found that the best response when you slip up on your diet is not to yell at yourself or give up and say you'll "start again on Monday."[5] When you slip up, forgive yourself, learn from it, and start again *now*.

Now you're armed with motivational tools to help you not only succeed but also enjoy your self-experimentation adventures. Next, it's time to pick mental targets. What aspect of your mental performance will you upgrade?

TAKEAWAYS

1. Choose your type of lab notebook carefully and use it daily.
2. Choose a compatible accountability buddy and practice good communication and support. Or start an accountability chain.
3. Share *relative* data, not raw data to maintain your privacy. Rather than sharing the actual numbers, this means sharing only the percentage that your mental performance has changed since you started your self-experiment. You compute the percent change relative to your baseline scores.
4. Use motivation hacks to keep going, including implementation plans and habit stacking.

Chapter 6

Debugging Yourself

"Greater in battle than the man who would conquer a thousand-thousand men, is he who would conquer just one—himself."
—Buddha

Time Investment: 25 minutes
Goal: To identify lifestyle and health challenges
that can cause mental inefficiency

Throughout my life, my mental energy levels have vacillated, sometimes dramatically. One afternoon during middle school, my math teacher waited impatiently for me to solve an easy multiplication problem. It was taking me a while, because my head was full of cotton balls that day. "Not exactly MIT material here, are we, Elizabeth?" A decade or so later, walking up to the podium to receive my MIT diploma, I chuckled at the idea of emailing a photo of my degree to that math teacher. I didn't, but still, that comment triggered something. I remember walking away from my math class that afternoon vowing that I would figure out what was causing my brain's random slowdowns. Little did I know there was no single root cause. Instead, there were many. As you begin debugging yourself, you may discover the same is true for you. If so, be patient and embrace the journey of self-discovery—in all its twists and turns.

A few years after my middle school math humiliation, I had just finished sprinting my final lap around a track. My ribs felt like flaming vises, my throat was closing, and my breathing was coming in an odd jerky rhythm. I sat down next to the track, but it didn't help. My coach came over. "Where is your inhaler?" I looked up, confused and gasping. "You are having an asthma attack. Where's your inhaler?" I didn't have one—because I didn't know I had asthma. After a peak flow meter test indicated that air was not moving out of my lungs well, my doctor put me on medication and gave me an inhaler. A happy side effect? On my new medicines, I had more mental as well as physical energy. As asthma can be difficult to spot, people with fatigue-related mental performance issues can test themselves with a peak flow meter. They're roughly $20 devices, and you can buy them without a prescription.

As a result of the asthma discovery, my doctor sent me for allergy tests (news to me: people with asthma often have allergies, and vice versa).[1] Time for a skin test: small metal teeth coated with potential allergens gently shoved into the skin on my forearms. Within minutes, large red welts had formed on my skin. Among other allergens, I discovered I was allergic to dust mites. The immediate solution was to improve the place I spent the most time exposed to dust mites: in bed. I purchased an air purifier and pillow covers. But a more long-term solution came out years later. The ear, nose, and throat doctor I went to told me about a wonderful innovation: I could get desensitized to dust mites and my other allergies! This FDA-approved treatment of sublingual drops (liquid that goes under the tongue) would allow my body to actually build a tolerance to these allergens. After I started the treatment, I could sleep in a hotel without carrying my own pillowcases or worrying about awakening to the sight of massive bags under my eyes and a puffed-up face. Best of all, the brain fog that I always had for the first few hours after waking has, over the course of my sublingual immunotherapy treatment, slowly gone away. If your mental performance suffers from low energy levels, it's worth investigating how certain elements of your environment (for me, dust mites and other allergens) may be affecting you and what steps you can take to make it better (allergy tests and, if appropriate, sublingual immunotherapy).

Physical Health

Before you pick your target for neurohacking, it's worth starting your journey by looking at your general health and lifestyle habits. If you can improve conditions for yourself in any of these areas, not only will your quality of life improve, you will likely find that your overall energy and attention will, too.

I've personally upgraded myself many times over both by treating medical problems (more on this below) and by upgrading my health and lifestyle.

- **Sleep:** Many students experience the mind-altering misery of an all-nighter at some point during their school years. New parents know how sustained sleep debt—at least until the baby learns to sleep through the night—alters mood and makes even the smallest decisions incredibly challenging. In addition to degraded alertness and vigilance, aspects of higher-order cognition that seem to be particularly vulnerable to the detrimental effects of sleep deficit include creativity, innovation, and emotion processing.[2]
- **Air quality:** I live in San Francisco, so during the California fire season or when I cook indoors and can't open the windows, the indoor levels of air pollution can skyrocket very fast. Even short-term exposure has been shown to lower mental performance.[3]
- **Temperature:** Being too hot or too cold can affect mental performance in surprisingly acute ways. Frequent heat exposure, coupled with inadequate air-conditioning, is linked to lower test scores and reduced learning.[4] Conversely, I personally suffered multiple weeks of lowered productivity during one Boston winter spent working in an old, unheated house.
- **Nutrition:** Many Westerners do not eat enough fruits, vegetables, nuts, and seeds. Being deficient in the micronutrients that these foods impart can lead to low mental energy, brain fog, headaches, and difficulty paying attention.[5] Additionally, many of us are sensitive to gluten, dairy, and caffeine. This can lead to inflammation and decreased mental performance as well.[6]

TREATING NUTRITIONAL DEFICIENCIES

Certain nutritional deficiencies can cause brain fog, lethargy, difficulties with paying attention, impaired learning and memory, headaches, and more. For children, certain nutritional deficiencies are associated with impaired cognitive development.[7] As of 2012, the CDC estimated that 10 percent of Americans have a vitamin or mineral deficiency.[8] The most common deficiencies in the US were, in order from most to least: vitamin B6, iron (in women 12 to 49 years old), vitamin D, iron (in children 1 to 5 years old), vitamin C, and vitamin B12. Far fewer people had deficiencies in vitamin A, vitamin E, and folate.[9] To date, I've been both anemic (iron-deficient) and severely vitamin D–deficient. My iron deficiency occurred because I was relying heavily on fortified cereal as my primary source of iron, which I consistently consumed with a calcium-rich food: milk. Unbeknownst to me, a number of studies have found that calcium inhibits the body's absorption of iron,[10] so my body was absorbing only a fraction of the iron I was consuming in my cereal. I had to begin to integrate other sources of iron into my daily intake.

Many popular diets that involve cutting out certain food groups come with increased risks of nutritional deficits. If you're vegetarian or vegan, gluten-free, low-carb, paleo, or keto, you're at a higher risk for deficiencies. For example, creatine—the stuff you may have heard weight lifters use to swell their muscle size and increase their power output—plays an important role in the ability to generate adenosine triphosphate, which make a difference when your muscles *or* your brain are working extra hard. Creatine is naturally found in red meat, so it can be hard to come by in a vegetarian or vegan diet. Studies have shown that taking creatine supplements helped vegetarians enhance their reasoning ability and/or working memory significantly— possibly because their diet had created a deficiency.[11] Older adults tend to consume less animal protein and should be especially vigilant of creatine levels.[12]

No need to stop doing what you're doing if your diet is working for you, but it's worth checking your blood levels for key vitamins and minerals periodically. In my case, I experimented with vegetarianism for a year when I was in college. Unfortunately, I didn't simultaneously experiment with learning about nutrition. Within months, I was feeling sluggish and slowed-down; it turned out I had developed an iron deficiency. Although vegetarianism is a healthy lifestyle for many, after my blood test, my doctor recommended I return to my previous diet since it was clear to both of us that I couldn't be trusted to eat sufficient amounts of lentils or tofu to make up for the lack of meat.

A few years out of college, when my roommate overheard me talking about my lifelong, periodic bouts of excruciating stomach pain, she kindly suggested I try a gluten-free diet. I not so kindly informed her that I didn't feel like eating rice and sawdust-flavored garbage for the rest of my life, thanks anyway. She promptly baked a wildly delicious batch of gluten-free cornbread and cookies. Touché. I agreed to a multi-week experiment of minimized gluten. By the end of the experiment, my rate of flare-ups had reduced considerably, and my mind felt surprisingly clearer. Giving up most of my previous main sources of carbs in exchange for a massive reduction in my usual intestinal agony was a no-brainer. I try to get blood tests every year for the vitamins and minerals already mentioned.

Prevalent Nutritional Deficits During Restrictive Diets[13]

Diet	Main nutrients of concern	Likely cause of deficits		Solution	
Gluten-free	Calcium Iron Magnesium Zinc — Folate Thiamin Vitamin B$_{12}$ Vitamin D	Many wheat products are fortified with nutrients	Possible over-reliance on rice-based products	Careful dietary planning with nutritious gluten-free foods	Fortified rice products
Vegan	Calcium Iron Zinc — Vitamin B$_{12}$	Meat and dairy are no longer sources of nutrients		Careful dietary planning with nutritious vegan foods / Supplementation with Vitamin B$_{12}$ — Pumpkin seeds for zinc / Spinach for calcium and iron	
Low-carb (e.g., Atkins)	Calcium Copper Magnesium Potassium — Pantothenic acid Vitamin E	Many nutritious foods are avoided	Many wheat products are fortified with nutrients	Careful dietary planning with an emphasis on low-carbohydrate vegetables and nuts	
Paleo	Calcium Iodine — Riboflavin Thiamin	Many nutritious foods are avoided		Careful dietary planning with an emphasis on leafy greens and nuts	

TRACKING YOUR VITAMINS AND MINERALS

Blood tests will provide certainty, but you can use a food tracker between tests. Most diet trackers look at your macros (the ratio of carbs to protein to fat that you're taking in), but a few also look at micronutrients (how much you are getting of key vitamins and minerals like magnesium, choline, vitamin D, and B vitamins). The value of tracking micronutrients is that you can notice patterns and detect a deficiency sooner than your next blood test.

You'll need a diet tracker with a comprehensive database of foods with complete nutritional data and, so that you'll use it reliably, look for one that makes it easy to enter data. I like apps where you can take a photo of the barcode on the food item or enter the food name in a mini search engine. The app I use to track my micronutrients is called Cronometer (it's available on both Apple and Android). Track for at least a week but ideally for a few weeks. Your data may help convince your doctor that it's worth getting a blood test. If your doctor is still not game to test you or if your insurance doesn't cover these tests, there are third-party test companies that you can order from directly as a consumer, too.

FOODS OR SUPPLEMENTS?

In comparison with taking a vitamin to fix a nutritional deficiency, it's usually more effective to just eat more foods that contain the nutrient you're deficient in. For reasons that we don't fully understand, the body seems to be better at absorbing nutrients from food than from multivitamins.[14] However, there are some caveats. If you're severely deficient, it can be hard to consume enough through food to make up for your deficiency; for example, when I was severely vitamin D–deficient, my doctor prescribed a pill that gave me roughly forty times the daily requirement until my levels recovered. It would have been very hard, possibly dangerous, to get that amount of vitamin D from food or sunlight as quickly. When selecting vitamins, minerals, or supplements, I highly recommend checking out each product's purity levels through third-party watchdog and independent testing groups such as Labdoor and Consumer Labs.

ARE EXTRA VITAMINS AND MINERALS EXTRA HELPFUL?

Let's say that you have no deficiencies. You're a model eater. Should you take vitamins? Not necessarily. Some of the vitamins most tied to cognition — the B vitamins (including B6, B12, and folate), for instance — are water soluble, so consuming more than you need typically leads to simply peeing them out, although high intake has been linked to some health issues.[15] While correcting deficiencies could provide significant cognitive boosts, the body of research suggests that taking extra vitamins provides little to no discernible cognitive gains for middle-aged and older adults who do not already have deficiencies.[16] For adults in the general population, there's only limited evidence that omega-3s enhance cognition.[17] There is mixed evidence that taking omega-3s while pregnant can improve the baby's cognition, and it probably doesn't hurt.[18]

Furthermore, a slew of studies over the last decade showed that people who took daily multivitamins did no better at avoiding all-cause mortality (that is, avoiding death from common causes such as heart attack, stroke, and so on) compared with those who didn't take multis.[19] One study found that cases of taking supplements of certain minerals (iron and copper) was associated with a slight increased risk of death over the long term; however, the study did not account for the possibility that some supplements may have been taken for reasonable cause in response to symptoms or clinical disease.[20] Furthermore, taking greater doses of supplemental vitamins and minerals may increase the risk of exposing yourself to heavy metals or contaminants.[21]

MELATONIN

While it does not treat a nutritional deficiency, melatonin may help treat (or at least manage) some sleep problems fairly safely.[22] If you can't fall asleep at night, using between 0.5 and 15 mg could do the trick.[23] Melatonin seems to be safe for short-term use, but its effects are unknown for long-term use.[24] Sleeping medications, however, present more complications. Sleeping pills induce a state that is closer to being sedated or anesthetized—not true sleep. They can also be habit-forming.[25] Furthermore, they seem to disrupt your brain's ability to sleep naturally; this will ultimately make you more tired and disrupt many of the processes that occur during sleep—learning, memory, attention, and emotional regulation.[26]

TREATING MEDICAL CONDITIONS

A few years ago, I got a lucky brain upgrade from an unexpected source. I had been experiencing persistent stomach pain and brain fog, and I'd hopped from specialist to specialist with no insights. By chance, my regular doctor wasn't available for a routine visit, so I got assigned to a new doctor—an MD who also happened to have a PhD in infectious diseases. I had just about given up asking doctors about my recent resurgence of stomach and brain fog issues, but when he asked whether I had "any other complaints," I decided to go for it. Immediately, he asked whether I'd been tested for a bacteria called *H. pylori.* I'd never heard of it. As he predicted, my test results came back positive. The doctor explained that while many people carry this bug and don't exhibit symptoms, a recent trip abroad that involved some adventurous eating may have given me enough to trigger my symptoms. After an intense treatment of antibiotic pills (and probiotic pills afterward to repopulate my gut), my stomach and brain fog symptoms abated. Had the *H. pylori* really gone? The stool test—which involved pooping into a pail and then having doctors examine my excrement—came back negative. Voilà! Bugs gone, brain upgraded. While this is a very young field, with much more research to be done, if you have been experiencing intermittent symptoms of brain fog, headaches, and even mood swings, it may be worth talking to your doctor about your gut microbiome health and, possibly, getting tested for an underlying infection.[27]

Mental Health

Addressing mental health challenges can improve your cognition massively. Mental health is fundamentally critical to mental performance. During the coronavirus pandemic, people became acutely aware of social and emotional connection in a world where we couldn't reach out and touch each other easily or safely. Feeling connected to others or to something larger than yourself is correlated with positive motivation, which also affects your performance.[28] In our increasingly online world, loneliness is on the rise. Loneliness is associated with serious health outcomes such as depression, suicidality, and poorer general health.[29] The cognitive consequences of major depressive disorder (MDD), which affects about one in fourteen American adults,[30] can be significant. There are a number of structural and functional alterations in the prefrontal and limbic areas of the brain, regions implicated in emotional regulation, in patients with MDD relative to healthy controls.[31] Depressed people experience drops in their processing speed, issues with working memory, and drops in their overall IQ.[32] Thankfully, these impairments don't have to be permanent. Multiple large-scale studies have shown that with effective treatment, individuals who had previously suffered from depression can recover their cognitive functioning.[33]

While estimates of mental health disorders vary across the world, one in five people in the US will be affected by some form of mental illness during their lifetime.[34] This ranges from common ailments like anxiety and depression to the relatively rare, such as schizophrenia. Mental health issues often start early and tend not to get treated for a long time; by age 14, around 50 percent of the people who will go on to a later diagnosis of a mental health condition are already showing signs, and by 24, it's closer to 75 percent.[35] On average, it takes people 11 years to actually get the help they need. In addition to all the other unpleasant symptoms and side effects associated with mental health struggles, that likely means 11 years of suboptimal cognitive functioning, too.

There are many different mental health struggles that affect mental performance, but, thanks to researchers, clinicians, and other healthcare workers, treatments are becoming increasingly effective and accessible each year. Just like an annual exam or a yearly flu shot, I take some type of mental health screener every year. I recommend you do, too. Not sure whether your mental or neurological health is optimal? Perhaps you feel a bit off but can't quite put your finger on the issue? It's easy, just check: take a general mental health battery and, depending on what you find or if you're unsure of the results, schedule an appointment with a psychiatrist or psychologist. There are ways of doing this in private without leaving a digital trail; for instance, if you use the privacy mode in your browser, you can take the online mental health screening at nami.org, the website for a grassroots mental health organization called the National Alliance for Mental Illness.

Depending on how you were raised or the attitudes of people you spend time with, the idea of seeing a mental health professional may make you feel uncomfortable. If that's the case, there are alternative ways to get help that better protect your privacy. If your appointment is via video conference, phone, or through an app (such as BetterHelp, Talkspace, Doctor on Demand, and others),[36] you run zero risk of someone recognizing you in the waiting room of a doctor's office.

I can't emphasize enough that if you're hoping for a cognitive boost, and if you discover that you have a mental health issue, treating it (with talk therapy, medication, or a combination) could be life-changing *and* cognitively enhancing. Once your health is better managed and you are in recovery, you may even see measurable cognitive test performance improvements.

Now, let's hunt through your health, lifestyle, and productivity to see if there are any bottlenecks hampering your mental performance.

Self-Assessment of Health and Lifestyle

In the following survey, you will assess ten areas of your health and lifestyle that could be in need of optimizing. Any of these areas could be bottlenecks blocking access to your peak mental performance. You will also answer questions about recent life events that could affect your mental performance; these questions include a mix of positive and negative life events. At the end of the survey, I've included a brief set of suggestions to try if you discover potential bottlenecks. I highly recommend that you retake this survey before and after you run any self-experiments aimed at improving mental performance.

Remember, you are not competing with anyone else. These questions provide a way for *you* to gauge where *you* think you're at on your health and lifestyle for the time frame specified. Just answer as honestly as you can. Grab a pen or pencil: it's time to reflect!

Health & Lifestyle Survey: Reflecting on the Recent Past	
PICK YOUR TIME FRAME OF REFLECTION	
Answer **all questions below thinking about a typical day** over the last 30 days or the last 3 months. Pick whichever time frame is most similar to how you expect the next 30 days to be.	
LAST 30 DAYS ___ LAST 3 MONTHS ___	
For the following questions, answer from 1 to 5 or n/a:	

1   practically never: 0–10% of the time
    2   sometimes: 11–35% of the time
    3   about half the time: 36–65% of the time
    4   most of the time: 66–90% of the time
    5   almost always: 91–100% of the time
 n/a I don't know or it doesn't apply.	

1. SLEEP	
I woke up feeling refreshed.	
I got just the right amount of sleep for me.	
I felt alert throughout the day.	

2. HYDRATION	
I didn't feel thirsty.	
I urinated a few times a day.	
My urine was clear or nearly clear.	
3. AIR/BREATHING	
I was able to breathe easily; I felt no constriction in my lungs, nose, mouth, or throat.	
I felt energized by my breath; I was taking in enough high-quality air each time I breathed.	
The air was easy to breathe; it did not smell odd or unpleasant.	
4. NUTRITION	
I did not feel distractingly full or hungry.	
It didn't hurt to poop.	
My nutrition was balanced across micronutrients (I got sufficient vitamin D, choline, omega-3s, etc.).	
I ate the right ratio of macronutrients (carbs, proteins, fats).	
5. PHYSICAL STRENGTH/ENERGY	
I got at least 30 minutes or more of exercise (an activity where your heart rate picked up, you breathed heavily, you exerted effort, and/or you engaged in sustained physical movements that were challenging for you).	
I had the strength or energy to keep up with the physical demands of my day; I didn't feel winded while performing routine parts of my daily life (climbing stairs, catching a bus, walking the dog, carrying grocery bags).	
6. SAFETY	
I felt safe physically (from other people, from the physical environment, etc.).	
I felt safe emotionally (from other people, from the physical environment, etc.).	
7. SOCIAL CONNECTEDNESS	
I felt connected to other people.	
8. SPIRITUALITY/MEANING	
My life felt meaningful and positive; I felt a sense of purpose; I felt connected to a bigger picture.	
9. MEDICAL HEALTH	
I did not have any health condition that required ongoing medical care. I did not have any other conditions that caused me to lose time at work, school, or other activities.	

I did not have any psychological or neurological conditions diagnosed by a neurologist, psychiatrist, psychologist, or other licensed mental health provider. I also did not suspect that I have an undiagnosed psychological or neurological condition that is affecting my daily functioning.	
10. LIFE EVENTS & STRESSORS	
For the remaining questions in this survey, the scale is different from the questions above. Answer the following statements on a scale from 1 to 5, where 1 is agree and 5 is disagree.	
Birth & children	
I was trying to have a child or was preparing to have a child or I recently had a child, or I just got a new sibling.	
Relationships	
I recently started a new romantic or domestic relationship.	
I was preparing to get married or recently got married. I (or someone in my immediate family) recently underwent a divorce or end of an important relationship (romantic or otherwise).	
Trauma	
I or someone close to me recently had an accident, got imprisoned, or required major hospitalization. Something else that recently occurred comes to mind when I hear the word "trauma."	
Death	
Someone close to me died recently.	
Change in economic, occupational, or social situation	
I recently lost a job, got a new job, began at a new school, or started a new semester.	
My financial status recently changed (gain or loss).	

Scoring the Health & Lifestyle Survey

Read through your answers to the questions above. If you have a high score in a given area, you likely do not have a bottleneck in that area. Lower scores indicate you may have a bottleneck in that area that may be contributing to lowered mental performance.

Using the Health & Lifestyle Survey for Self-Tracking

To get a more rigorous baseline, or to use the above questions as part of a self-experiment, you can take the following approach:

Do not use the 1–5 scale indicated in the survey above. Instead, record a "yes" or "no" in response to each of the questions in the above assessment every night before bed.

Your score for each category (sleep, hydration, air, etc.) will be based on the number of nights you answered "yes" to each of the questions in that category divided by the total you could have had if you'd answered "yes" to every question every night.

Let's break that down. Let's say you gathered data for seven days. If you had answered "yes" to every question in the sleep category every night, you would have had a total of 21 (three questions times seven days). However, let's say you answered yes to only two of the sleep questions for four days. Your score would be eight (two "yes" answers times four days) divided by 21 — or, roughly 38 percent.

SUGGESTIONS FOR HEALTH AND LIFESTYLE BOTTLENECKS

Each of the topics in the survey could be the subject of its own book, so I will not pretend to have all the answers in the small space we have here. I encourage you to find books and resources specifically devoted to any of the domains where you identify a bottleneck. That being said, here are a few solutions to problems above that I've come across and found to be particularly insightful or useful.

Sleep

Possible bottlenecks: Make a note of any poor sleep habits, sleep environment, suboptimal duration, or timing of sleep issues.

Possible solutions: To prevent the computer or phone's blue light from disrupting your sleep schedule, use dimming software like Night Shift on Mac or Android's Blue Light Filter. If you suspect that you have mild sleep apnea, a surprising treatment — from a randomized controlled trial published in the *British Medical Journal*[37] — is to learn how to play the long, wooden Australian didgeridoo. Other wind instruments may help, too. The theory is that improving strength and control in the throat helps treat the underlying cause of sleep apnea. Naps can be a good sleep habit

---

to develop, too. Power naps of even 6 minutes improved declarative memory (the kind you use when memorizing facts).[38] The US military perfected an approach to falling asleep that, if practiced repeatedly, apparently allows you to fall asleep within 120 seconds. It involves progressively relaxing each part of your body (face, then neck, then shoulders, then arms, and so on). It is similar to a meditative body scan.[39]

To prevent light and noise disruptions while you sleep, get an eye mask and ear plugs. To fall asleep faster, develop a pre-sleep routine (such as reading for 10 minutes).

Hydration

Possible bottlenecks: Hydration.

Possible solution: Carry a water bottle with you at all times and set reminders to drink until it becomes a habit. Check local tap water quality, and use a reusable water bottle made of glass, BPA-free plastic, stainless steel, or aluminum.[40]

Air/Breathing

Possible bottlenecks: Temperature, pollution, asthma, environmental allergens. The World Health Organization recently came out with a statement that nine out of ten people globally lives in bad air conditions. Four out of ten Americans are living in poor air conditions.[41]

Researchers have found that women typically work best under warmer conditions (around 77 degrees Fahrenheit) and men under cooler conditions (around 72 degrees Fahrenheit).[42]

An analysis of children's standardized test scores during heat waves, using economic models to account for dozens of different variables, found that for every one degree Fahrenheit increase above the recommended temperature in a testing center, American students typically lose 1 percent of their PSAT score.[43]

What about allergies? Constant sneezing and a swollen nose are not only annoying, researchers measured the cognitive performance of those afflicted; it plummets during pollen season.[44]

Possible solutions: Monitor the temperature with a thermometer and the air quality using an air quality monitor. Adjust the thermostat or use a room heater, fan, or air conditioner; use an air purifier. Consider going to a doctor to get checked for allergies and asthma if you are having difficulties breathing; if you have environmental allergies (such as ragweed or dust mites), look into sublingual or other types of immunotherapy that may improve your quality of life. Dust mite–blocking pillowcases and bedding may help reduce symptoms, too.

Nutrition

Possible bottlenecks: Total number of calories consumed, distribution of caloric intake throughout the day, blood sugar spikes, digestive issues, allergies; while these differ for everyone, some common culprits are: caffeine (including foods that contain caffeine, like chocolate), dairy, foods containing significant gluten, and foods with a high glycemic index (very sugary).

Possible solutions: Certain eating habits have good research behind them, such as the MIND diet. Experiment by keeping a food diary and eliminating potential trigger foods that give you brain fog or rob you of energy. If you are suffering from brain fog, consider trying a short elimination diet to see whether some of the common food triggers bother you, too.

Physical Strength/Bodily Energy & Lack of Pain

Possible bottlenecks: Weaknesses in muscle strength or cardiovascular capacity that drain physical energy, distracting pains or aches.

Possible solutions: If there is no clear medical cause, starting an exercise regimen to increase your strength and endurance could reduce the effort required for your daily life and provide an overall energy boost.

Safety

Possible bottlenecks: Specific places, situations, or people are distracting you and creating uncomfortable levels of stress that make it hard to focus, learn, and be creative.

Possible solutions: This is a deep and complex subject. Depending on your triggers, your response could be reaching out for support (the human resources department at your company or student support service at your school; help from a therapist, counselor, or mediator; asking for help from local political or religious leaders). Or, it could be developing new personal skills, habits, or mindsets (for example, using crime maps to change your route home to avoid known trouble spots,[45] taking a self-defense class, learning meditation, limiting your exposure to news and social media). Other solutions may require significant and extremely difficult changes in one's life (such as leaving an abusive relationship or a toxic workplace, moving neighborhoods, etc.). Pinpointing and implementing changes that work for you and are within your present means will not only help you create a safer environment for yourself, it will be critical to helping you maximize your daily mental functioning.

Social Connectedness

Possible bottlenecks: Feeling lonely is distracting, painful, and is tied to serious health conditions.[46]

Possible solutions: Social health is a critical aspect of mental health.[47] Helping others through volunteer work,[48] writing in a gratitude journal,[49] and cognitive behavioral therapy[50] have all been shown to help alleviate loneliness.

Meaning/Spirituality

Possible bottleneck: Feeling like your life is not meaningful can make it hard to concentrate, be creative, or make full use of your mind.

Possible solutions: Meditation, prayer, volunteering, spending time in nature, joining a spiritual community, or engaging in other activities that inspire feelings of awe, gratitude, and altruism.[51]

For you to get the most out of your self-experiments, you'll need to alleviate these bottlenecks. Once you treat them, you will likely experience a gain in your mental performance, not to mention your sense of well-being. Many books have been written on each of these areas, but this chapter lists some uncommon solutions and basic ways to get you started.

TAKEAWAYS

1. Assess ten key areas of your health and lifestyle (such as sleep, hydration, air quality) to discover potential bottlenecks contributing to mental performance issues. Survey recent events in your life for potential stressors, as these can affect mental performance, too.
2. Retake the health and lifestyle survey, as well as the life events survey, periodically, especially before and after you run any self-experiments aimed at improving mental performance.

In the next four chapters, we'll consider four domains of mental functioning that you may want to consider targeting for an upgrade. The first is executive function, which plays a key role in how we learn, work, and manage our daily lives.

PART II

Find Your (Mental) Focus

Chapter 7

The New IQ[1]

Time Investment: 23 minutes
Goal: To learn what executive function is and
test your own

Some conversations from high school stick with you. The girls in my dorm used to sit at a particular table in the dining hall, and a boy named Mark used to eat with us sometimes. One day I came to lunch late and sat down, trying to catch up with the conversation. Mark had presented an idea and dropped out of the conversation — watching and saying little as the rest of us debated it heatedly. Occasionally, he would ask a question that steered the conversation in a way that made it clear he had listened carefully, synthesized all our points, held back his own opinions lest they influence us, and was now eager to probe our thoughts in a new direction.

He also zoomed from the big picture of our discussion — in this case, the potential of the internet for social interaction — down to the details, remembering which of us had taken any particular stance over the course of the conversation. Three things stood out to me: how much he seemed to be able to hold in his mind at once, his ability to hold back his own opinions in order to avoid biasing others', and how flexibly he moved from one part of the conversation to the next. He seemed to watch our expressions, weigh our arguments, and do some kind of internal assessment — all quickly and, seemingly, effortlessly. Observing those three abilities led me to ask three questions. Did he have more room in his mind to fit all the information relevant to any particular moment, like an artist working on a bigger canvas? Did he have better self-control than the average kid? Could he shift mental gears more quickly?

If you haven't guessed already, Mark's last name was Zuckerberg. A few years later, he started Facebook. I have no idea whether that high school discussion had any effect on his entrepreneurial path, and my goal here is neither to endorse nor criticize Facebook. Rather, that long-ago discussion showcases three key aspects of a mental ability called executive functioning. The three aspects—how Mark seemed to have more room in his mind to hold information pertinent to the present moment, a strong ability to hold back his own opinions when necessary, and an impressive mental flexibility—all correspond to key aspects of executive function.

So what is executive functioning, exactly? And what can boosting it do for you?

The Science Behind Executive Functioning

Imagine the following: You can mentally accelerate so fast it's as if everyone else is half asleep. You can examine ideas with microscopic detail or leap multiple steps ahead in any sequence, thinking of all the potential outcomes, playing chess with possibilities before others even recognize that they exist. You can hold enormous amounts of information in your mind at the same time and manipulate it to solve problems in real time that would otherwise require you to write out everything first. If you could do all that, you'd be an executive function superhero.

Executive functions, as the name would imply, are the mental abilities that are in charge of other abilities in your brain. What do executives of the brain do all day? Like a CEO, they plan and make decisions, correct errors, and troubleshoot. They handle unexpected situations, and they monitor and jump in during difficult tasks that the rest of the brain does not yet know how to do. They pay attention to threats and opportunities in the outside world, try to organize and make sense of new information, and compare it to previous experiences.

If executive functioning is a CEO, that CEO relies on what I think of as the WIF team, the sub-abilities of working memory (W), inhibition (I), and mental flexibility (F).

Working memory is your ability to not only hold information in your mind but also manipulate it. It's exercised when you hear a phone number and have to use the digits that you've held in your mind to enter them into your phone, when you multiply large numbers without using paper and pencil, when someone suspects you weren't listening and asks you to repeat what they just said—and you do so.

Inhibition keeps you from blurting out the punch line of a joke, helps you ignore the cake and pick the salad when dieting, and keeps you focused on an important email instead of binge-watching cat videos.

Flexibility gives you the ability to shift your attention from one idea to another and to incorporate multiple ideas at once as needed. This can come up when you are switching from one email conversation to another, or when you are synthesizing ideas from multiple sources while writing up a report.

Taken together, executive function gives you sustained focus, goal orientation, ability to do tasks sequentially, ability to stay organized, and ability to tackle novel tasks.

Part of the executive function's role is to monitor your arousal level—how alert and engaged you are—and to keep you in a safe, effective range. Your arousal level is predictive of how well you'll be able to handle difficult tasks. Too much anxiety and you'll be able to complete only very simple tasks,[2] but if you're half asleep you'll be of no use on even those. If you want to succeed at difficult tasks, you'll need a medium amount of arousal—not too much, not too little.

Not Too Much, Not Too Little: Optimal Mental Performance at a Medium Arousal Level[3]

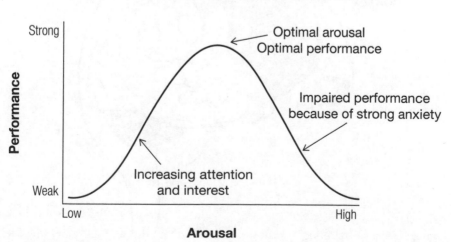

Where is executive function in your brain?

As I mentioned earlier, IQ testing is rife with cultural bias and thus is a questionable measure of mental performance. Executive functioning, on the other hand, is something that we do have ample neuroscientific data on, and is, in that way, far more *real*.

If you tap on your forehead, your executive functions live less than an inch below your fingertips (through skin and skull) in a region of the brain called the prefrontal cortex. Researchers learned these locations by watching what parts of the brain lit up when people lying in brain scanners made decisions, planned, and performed other executive functions.[4]

In order to use information to make decisions and plan, the frontal cortex has to talk to other parts of the brain where memories are stored and new sensations are processed. In addition, it has to tell the rest of the brain its decisions, which will initiate behaviors and actions. Since memories, sensations, and action initiation heavily involve other areas of the brain (not just the frontal area), a task involving executive functioning will typically involve multiple brain regions, too.[5]

Key Region for Executive Function: The Prefrontal Cortex[6]

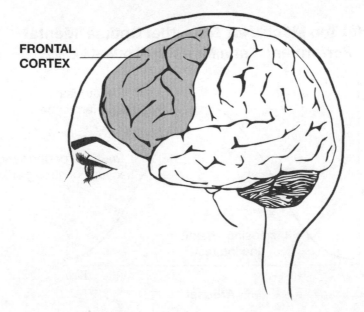

FRONTAL
CORTEX

Why Working Memory Is So Important

Executive function seems to be tied to success across various aspects of life. A cohort of over 1,000 children in a city in New Zealand was followed from birth to 32 years old.[7] Researchers looked at the children's levels of self-control, an ability related to executive function through reliance on inhibition, among other cognitive abilities. It turned out that self-control in the first ten years of life predicted the individuals' health, wealth, substance dependence, and whether or not they ended up with a criminal record. Even when adjusting for the wealth and status of the parents and the child's performance on IQ tests (as I've said, a non-ideal assessment, but one still used in a lot of research), the assessment of their self-control still strongly predicted their outcomes. Further corroborating the findings in New Zealand, multiple American studies found that students' academic performance from elementary school through university could be predicted by simply knowing their working memory scores as small children—in fact, predictions based on working memory were more accurate than those based on IQ.[8]

A host of conditions go along with executive dysfunction. It's probably not surprising that people diagnosed with ADHD, a condition characterized by difficulties with regulating attention, typically struggle with executive functioning. Less well known, however, is the fact that people with schizophrenia struggle with executive functioning, too.

The role of executive functioning in schizophrenia was brought home to me in a story told by researcher Christine Hooker, who at the time was at Harvard studying teenagers with early signs of schizophrenia. She told me that a teenage boy in one of her studies had woken up one day in his bedroom at home and smelled something oddly metallic. The walls appeared to be glistening with blood. He walked over to the wall and touched it with his fingertip. When he brought his finger to his nose, he smelled blood. Understandably, he felt absolutely terrified. At this point Dr. Hooker told me: "Then, he paused." He closed his eyes and asked himself whether it was likely that the walls *really* had blood pouring down them. Although all his senses were telling him that they did, he was able to hold all this information in his mind at the same time—and talk himself out of it. His logical reasoning kicked in. He decided the situation was making him too emotional to think straight, so he left his bedroom.

One of the things that was unusual about this young man compared to some of the other teenagers diagnosed with schizophrenia in her study, Hooker explained, was the size of his working memory. He could hold more information in his mind than the average person. She believed that was what helped him make sense of the confusing sensations and ultimately gave him the ability to overrule his senses to a degree that some of the others could not. In the years since, Hooker has been testing the efficacy of executive functioning training as a path for young people who struggle with psychoses to function more typically amid ongoing distraction from their unruly brains.[9] If someone struggling with as great a challenge as psychosis could manage it using greater working memory, imagine what the rest of us could do with access to more working memory.

Working memory also seems to help students succeed in school. Tracy and Ross Alloway, married researchers previously at the University of Stirling and Edinburgh respectively, found that working memory was a far better predictor than IQ for school performance. When they measured students' working memory in kindergarten, the Alloways were able to predict students' school performance in sixth grade with 95 percent accuracy.[10]

A recent study of child prodigies further illustrates the unique power of working memory within the executive functions. The children in the studies were defined as prodigies because, before the age of 10, they demonstrated the abilities of adult professionals in the fields of art, music, math, and other areas. Although many people believe that prodigies must have genius-level IQs, many did not. Their IQs ranged from 108 (slightly above average) to 147 (way above average), with some scores in the below-average range. While they excelled in a broad range of fields, six of the eight prodigies tested had a working memory score that was better than 99 percent of the general population.[11] A small observational study like this doesn't tell us whether their large working memories made them into prodigies, but it's plausible to imagine that it helped them master more material faster than they otherwise could have.

Executive function affects both your work and your personal life. How well you plan ahead when you have to go on vacation—and must juggle packing with finishing up work projects—comes down to the current health and functioning of your executive function. How well you hold multiple ideas in your head at once—say, when a colleague updates you on what they got done last week and how it will affect what you need to do this week — comes down to executive function, too. Think about how well you resist temptations when you're out with friends and one of them is pushing to order more than you are actually hungry for. Think about how often you come up with novel, creative solutions when your boss throws a problem at you that you've never seen before. In all of these situations, you are relying on your executive function.

Testing Your Executive Function

Testing executive function can be tricky. There are both self-report and performance-based tests of executive functions, and to get the most useful picture you'll want to use both. To get your self-reported baseline, you can use the self-assessment at the end of this chapter. To establish your performance-based baseline, you can use the tests at the end of this chapter; computer-based versions can be found on the web. Once you've tried the interventions in the rest of the book, go back and retest your executive function to see how much it has changed.

IS IT OK TO DO IT YOURSELF?

You may be wondering whether it's OK to use these tests at home instead of going to a psychologist's office. Assuming you are not looking for a clinical diagnosis, the answer is yes. While having a one-on-one test with a psychologist could be extra valuable because they could customize the tests based on their observation and interactions with you, the real value in neurohacking is in taking tests repeatedly — even daily! — and watching your changes over time. Furthermore, there is now evidence that some computer-based tests can be quite accurate, too. As far back as 2010, Laura Germine at McLean (a Harvard-affiliated mental health hospital) and colleagues showed that valid results are possible from computer-based tests at home.[12] More recent studies show the efficacy of online tests conducted at home as well.[13]

Self-Assessment of Executive Function[14]

This survey examines the core components of executive function — working memory, inhibition, and flexibility — as well as sustained focus, goal orientation, ability to do tasks sequentially, ability to stay organized, and ability to tackle novel tasks.

Remember, you are not competing with anyone else. These questions provide a way for *you* to gauge where *you* think you're at on your executive functioning for the time frame specified. Just answer as honestly as you can. Grab a pen or pencil: it's time to reflect!

PICK YOUR TIME FRAME OF REFLECTION

For this survey, please answer all questions with a specific time frame in mind.

If you want to use your answers to this survey for assessing the effects of specific interventions during your daily self-experiments, choose "Last 24 hours." Or reflect every 24 hours if you want a more accurate baseline.

If you want to use it as a way to record your more general sense of your executive function over a longer time period, choose "Last 30 days" or "Last 3 months."

Indicate which time frame you choose by checking one of the options below.

LAST 24 HOURS ___ LAST 30 DAYS ___ LAST 3 MONTHS ___

For the next set of questions, you will be evaluating your executive functioning on a scale from 1 to 5 (where 1 is poor, 2 is fair, 3 is medium, 4 is good, and 5 is excellent) in the following three contexts:

— Primary responsibility or focus: Pick the activity where achievement matters most to you. This could be work or school or another area.

— Personal projects: Evaluate whatever personal project matters most to you right now. This could be learning a musical instrument, planting a garden, volunteering in your community, etc.

— Life-maintenance activities and responsibilities: Evaluate how you are at remembering to eat, sleep, exercise, show up at the time you said you would, complete your chores, etc.

Your score for each question overall is the sum of your answers to each of the three parts. For example, if your answer to the primary responsibility part was 3 (medium), to the personal projects part was 2 (fair), and to the life maintenance part was 4 (good), your score for the question overall would be 9.

1. FOCUS AND GOAL-DIRECTION: How good were you at being focused and goal directed while engaged in the following activities?

Primary responsibility (e.g., work or school)	
Personal projects	
Life-maintenance activities	

---

2. SERIAL PROCESSING: How good were you at planning and taking steps one after another (i.e., planning and executing subgoals for a larger goal) in the below activities? This includes processing tasks sequentially, being able to break down goals into tasks, plan tasks in a particular order, execute on goals and tasks in planned order).	
Primary responsibility (e.g., work or school)	
Personal projects	
Life-maintenance activities	
3. CHANNELING ATTENTION: How good were you at controlling and sustaining your attention (ignoring distractions, focusing for long periods when the task must be done but is not particularly interesting or when you are tired, noticing when the situation has changed and adjusting what you are doing as needed) in the following activities?	
Primary responsibility (e.g., work or school)	
Personal projects	
Life-maintenance activities	
4. SOLVING UNFAMILIAR PROBLEMS: How good were you at solving problems you've never seen before in the following areas?	
Primary responsibility (e.g., work or school)	
Personal projects	
Life-maintenance activities	
5. ORGANIZATION: How good were you at keeping yourself neat and organized in the following activities?	
Primary responsibility (e.g., work or school)	
Personal projects	
Life-maintenance activities	
6. FLEXIBILITY: How good were you at being mentally flexible (able to change the course of your thoughts or actions based on a change in the situation) in the following activities?	
Primary responsibility (e.g., work or school)	
Personal projects	
Life-maintenance activities	

Scoring

Sub-Scores:

It's easy to focus on the overall score, but often the sub-scores of a quiz like this are actually the most useful. As you run your self-experiments, you may discover that a specific subset of these questions is the most interesting. For instance, it may be that one aspect of your life—your primary responsibility, personal projects, or your life-maintenance activities—is more of a bottleneck than other areas. Or, perhaps you sense that Organization—across all aspects of your life—is your biggest bottleneck (e.g., say you got a 10 in Organization but got a 12 in all other areas). Whatever subset feels right to you, I encourage you to pay special attention to that personally chosen subset during your self-experiments going forward.

Overall Scoring Range: 18–90.
Your final score is the sum of your answers to all the questions. What tier is your score in?

> Bottom third score: 18–41
> Middle: 42–66
> Top third score: 67–90

Lower scores indicate that executive function may be a bottleneck for you. If this is the case, it could be worth experimenting with interventions that target executive function. More in Part III.

Performance-Based Tests

The following are performance-based tests. They can be used daily before and after your intervention. Your scores should be recorded in your neurohacker's lab notebook. Note that all of the tests below are subject to practice effects—you get better at the tests by doing them more. For that reason, take your chosen test multiple times (say, once a day during your baseline week) and take the second-best score you get as your baseline score. Then, you can compare the scores you get during the intervention period to that baseline in order to see how much the intervention affected you (hopefully, it helped!). Note that the tests below are modified for use in a book, but you can find computer-based versions on the web.

Working Memory Test: Book n-Back

This is a test of working memory. Versions of it have been used since the late 1950s.[15] The following version has a more personal story, though. One night,

I was worrying aloud about writing this part of the book. "What if the reader is in a park or on a plane without internet?" My husband, a computer scientist, responded, "What if the reader just used the book itself?" My eyes bugged out. "Brilliant!" I exclaimed. That's how this version of the test was conceived.

Materials

- Timer
- Something to write with
- A book (any book—even this one!—with page numbers and letters on the pages)
- Neurohacker's lab notebook to record your session information and results

Directions

1. Open to a random page of the book. Write down the page number on your results sheet.
2. Start the timer for 1 minute.
3. Start with $n = 1$ to warm up. $N = 1$ means you remember the first letter from one page back; let's say it was "o." Go to the next page. Once you see a new first letter at the top of the page (let's say it's "l"), write down your memory of the first letter on the previous page. You should have written down "o" so far.
4. Look at the next page. As soon as you see the first letter on this page (let's say it's "p"), record the first letter from the previous page. So far, you should have written down two letters: "o" and "l."
5. Repeat until the timer runs out. This task is tricky for most people, so don't be discouraged if you only record about a handful in a minute.
6. If $n = 1$ is too easy, try $n = 2$ (meaning you'll have to record the first letter from two pages back but also remember the letter on the page one back and the one on the current page for future use), then $n = 3$ (meaning you'll have to record the first letter from three pages back but also remember the letter on the page two pages back, one page back, and the one on the current page for future use), and so on.

Scoring

Range: Give yourself an accuracy and a speed score. Accuracy is the percentage of correct letters you recalled out of the total number of pages you attempted. Speed is the total number of pages you attempted. The lowest score you could get is 0. The highest score would be the maximum number of pages you can flip through in one minute.

Processing Speed: The Ruler Drop Test[16]

This is a test of motor cognitive processing speed. Executive function is composed of more than just how fast you process information and take action—speed is one of the things researchers measure when they attempt to estimate executive function. One way to measure yours is to use a ruler to test your reaction time. While this is not purely a test of executive function, when you score well on processing speed, your executive function tends to be higher, too.[17] Scores can range from catching the very end of the ruler to catching it within a few centimeters.

Materials

- Ruler (with centimeters)
- Something to write with
- Neurohacker's lab notebook to record your session information and results

Directions

1. Have a friend hold a ruler above your hand with the "0 cm" mark closest to your hand.
2. Have them drop the ruler between your extended pointer finger and thumb of your dominant hand; your job is to catch the ruler as quickly as you can.
3. How many centimeters did it fall before you caught it? That measurement can be converted into your response score. For example, if it fell 10 cm before you caught it, then 10 cm would be your response score.

Make sure to repeat it multiple times (at least 5) and then take the average of your scores.

Additional Performance-Based Tests: Flanker-Like Test, Stroop-Like Test

Note that the Flanker and Stroop tests below are single examples. Once you've taken the ones below, you can't retake them, as you will have practice effects. Due to space constraints, I can't give you a full example here, but you will find more on the web; computer-based versions tend to be more similar to the ones used by researchers, too.

Inhibition Test: Flanker-Like Test[18]

This is a test of inhibition but with a twist. It is a classic neuropsychological task that has been used for decades, at least since the 1970s.

Materials

1. Timer
2. Written test below (for more examples, go to the web)
3. Something to write with
4. Neurohacker's lab notebook to record your session information and results

Directions

1. Write the correct direction ("r" for right-facing and "l" for left-facing) of the target arrow in question. Note the middle arrow is equidistant between the left- and the right-most arrows.
2. Example of right-facing middle arrow: < < > > > Answer: r
3. Set your timer for 1 minute. Answer as many as you can.
4. Scoring (see chart below): Give yourself 1 point for every correct answer, 0 points for unanswered or incorrect answers.

Question	Which direction is the middle arrow pointing?	Answer
1	> < > > >	r
2	< > > > >	r
	Which direction is the left-most arrow pointing?	
3	< < > > >	l
4	> < < > >	r

Inhibition and Flexibility Test: Stroop-Like Test[19]

This test assesses your ability to inhibit an automatic response and adapt to changing rules. It has been used for nearly a century, at least since the 1930s.[20] It assesses your ability to inhibit automatic responses and shift your perspective quickly. The name of a color is written in different-colored fonts with different background colors. Your job is to answer what the meaning of the word is or the color of the font, depending on which is being asked of you at the moment. For example, if you were asked what color is the font for Black, you should answer "white." If you were asked about the meaning, you would say "black."

Materials

- Timer
- Stroop prompts. Due to space, I've only included 10 prompts below, but for a minute-long test, I'd recommend you figure out the pattern below and prepare about 75 or so for yourself (or go to the web for full examples).
- Something to write with
- Neurohacker's lab notebook to record your session information and results

Directions

1. Set the timer for 1 minute.
2. Specify the color of the font.
   Example 1: Black Answer: W
3. Specify the meaning of the word.
   Example 2: Black Answer: B
4. Scoring (see chart below): 0 points for not answering a question or answering incorrectly, 1 or 2 points depending on the difficulty of the question.

	What color is the font?	Points	Answer to: What color is the font?
1	Black ___	1	B
2	White ___	1	W
3	Black ___	2	W
4	White ___	2	B
5	Black ___	1	B
	What color is the meaning?		Answer to: What color is the meaning?
6	White ___	1	W
7	Black ___	2	B
8	Black ___	2	B
9	White ___	1	W
10	Black ___	2	B

Executive function is one possible target for neurohacking improvement attempts. Now let's consider the next option: emotional self-regulation.

TAKEAWAYS

1. Executive function is composed of three mental abilities: working memory, inhibition, and flexibility. The kinds of activities for which you need executive function include planning ahead, holding multiple ideas in your head at once, resisting temptations, and coming up with creative solutions to problems. The part of the brain that seems to manage these abilities is the prefrontal cortex.
2. High executive function predicts success academically, professionally, and personally.
3. Low executive function is associated with struggling in school and work, including being more likely to drop out of school and earn less. It can also exacerbate mental health and neurological challenges such as schizophrenia, ADHD, and other conditions.
4. You can measure your executive function with computer-based tests (these may look familiar if you play brain games), as well as pen-and-paper tests.

Chapter 8

The New EQ

"Between stimulus and response, there is a space. In that space is our power to choose our response. In our response lies our growth and our freedom."
—Viktor Frankl

Time Investment: 12 minutes
Goal: To learn what emotional self-regulation is and test your own

---

Viktor Frankl is one of my all-time heroes. If I could host any 10 people alive or dead for a dinner, his invite would be one of the first. As a Jewish psychiatrist, he was sent to a concentration camp during World War II, but instead of giving in to despair and fear, he decided to make meaning out of the horror transpiring around him. He promised himself that he would pay attention to everything that was happening—to who was able to hold on to their humanity amidst all the suffering and who was not—so that, if he were lucky enough to survive the camps, he would be able to share with the world which psychological strategies had helped prisoners survive not just physically but also emotionally. Frankl survived, and he wrote up his observations in the massively bestselling book *Man's Search for Meaning*. Through his books and his private practice as a therapist using this set of perspectives and techniques, he helped millions of people discover a deeper sense of meaning and purpose in their lives.

I interpret the "space" from Frankl's quotation as a place to exert the "new EQ": emotional self-regulation. In that space between stimulus and response, we have a chance to make decisions. What we do in that space determines our future. Learning how to manage your emotions and thoughts during that critical space is what this chapter is all about, wherein lies "our growth and our freedom."

The Science Behind Emotional Self-Regulation, and Why It's So Important

Researchers and clinicians define emotional self-regulation as the ability to monitor, assess, and modify your own emotions. Your ability to modify your own emotions—how intensely you feel something, the variety of different feelings you experience, how long you feel particular emotions—is partly conscious, partly subconscious.[1] Self-control and willpower are related concepts, too. When you meet a person with strong emotional self-regulation, they may be full of contradictions. They may experience a broad range of emotions, but also demonstrate incredible mastery and flexibility in how and when they express them.[2] Emotional self-regulation doesn't mean being stone-cold.

The emotionally self-regulated person possesses two defining characteristics:

- **The ability to delay the expression of emotions:** If, for some reason, they are suddenly struck with the urge to laugh during a funeral, they control themselves. When the situation allows it, however, they are spontaneous with laughter, tears, or other emotions.
- **The ability to influence one's own feelings, thoughts, and physiology:** If you notice that your heart is racing before giving an important presentation and you interpret it as a sign of your impending death, causing your anxiety to spike even further, that would be an example of struggling with emotional self-regulation. Effective emotional self-regulation would be, instead, noticing your pre-presentation jitters and telling yourself, "That means I'm extra sharp and alert now—perfect!" That reframing would provide a calming and steadying influence to improve your performance.

High emotional self-regulation is tied to success in work,[3] school,[4] and relationships.[5] For instance, if you were in the final push preparing for a meeting with a client but you noticed that you were feeling stressed and couldn't concentrate, you might realize that, paradoxically, slogging on might waste more time than taking a break. You might get up, take a quick walk around the block, and then come back with renewed focus. Interestingly, a little anxiety seems to help us learn something well in the first place, but to remember it effectively, anxiety must be lower.[6] In the relationship category, emotional self-regulation reaps huge rewards. Being able to manage your own emotions provides a major benefit because you don't, for example, accidentally start fights by snapping at your partner when they ask an innocent question unrelated to the source of your stress.

Emotional self-regulation seems to take practice and time.[7] In study after study, older adults outperform younger adults and children at emotional self-regulation.[8] Relatedly, they also report higher amounts of positive moods and lower amounts of negative moods.[9] So what skills have they developed?

There are four main steps involved in effectively managing an emotionally charged situation. Successful emotional regulation involves recognizing each step as it happens and making mindful choices at each point.

- **Step 1 is Awareness:** realizing you are in an emotionally charged situation. This includes noticing your own physiology (e.g., realizing you are nervous because your hands feel clammy).
- **Step 2 is Selection:** choosing what you pay attention to within the situation.
- **Step 3 is Interpretation:** choosing how you interpret the situation.
- **Step 4 is Response:** choosing your response to the situation.

People with high emotional self-regulation are able to navigate situations more successfully than those who don't recognize or handle each step as carefully.

In the "Response" step, people high in emotional regulation seem to operate in a handful of ways.[10] Here are a few examples:

- They decide whether to stay in or leave emotionally charged situations.
- They find ways to lighten or ease the situation (for example, by winning over others through a self-deprecating joke).
- They may also redirect the focus of the situation, or change everyone's perception of the situation by telling a story.

- They may try to take on a different viewpoint to gain perspective on events.

- They may try to see themselves or their situation from another perspective. This could include taking on the perspective of the person they're in conflict with to build empathy, or building self-compassion by looking at themselves as they would a good friend.

- If the emotional stressor is long-term, they may cultivate new habits: scheduling a meditation class for the hour right before a demanding meeting or planning to take a refreshing nap after that presentation ends.

What's Happening in the Brain and Body During Emotional Self-Regulation?

Scans show us what is going on in the brain when it tries to emotionally regulate. When research participants in a brain scanner are asked to perform activities that involve emotional self-regulation—such as being provoked into strong emotions by watching a happy or sad movie—and asked to regulate their emotional responses, we often find that two sets of brain regions are highly active.[11] The first is the front of the brain (regions in the prefrontal cortex, the same part that is active in executive function tasks). The second set of areas that tend to be active, not surprisingly, are those involved in emotion generation, including the amygdala (which is involved in coding the importance or "salience" of things in our environment, and tends to be very active in instances of fear or threat). The strength of the connection between the frontal areas and the amygdala tends to increase when participants are asked to self-regulate their emotions as well.[12]

All this "lighting up" indicates that the brain regions involved in executive function and self-control are working to regulate activity in the emotional centers. People who struggle with emotional self-regulation tend to have markedly different brain activation patterns from those who find emotional self-regulation easier. People with higher self-control tended to have stronger connectivity between prefrontal regions and the amygdala.[13]

Emotional self-regulation is not all about the brain, though. If you've ever felt very angry, you may have noticed that multiple parts of your body felt hot. Your adrenal cortex (a region in your kidneys) sends fight-or-flight chemicals into your bloodstream, and that gives rise to a cascade of changes all over the body,[14] including increases in your fingertip temperature and dilation of the pupils in your eyes. Similarly, your heart rate variability (the rhythm and cadence of your heart rate)[15] and chemicals released by your skin all change in response to your emotions. Perhaps this accounts for why the types of meditation that involve focusing on parts of your body that get activated by emotion can be so successful in training people to better manage their emotions.

Testing Emotional Self-Regulation

Presently, we are stuck with surveys to self-report on your current level of emotional self-regulation. Over the next few years, you may see more biologically based assessments. They will likely detect differing levels of emotional self-regulation behavior and provide suggestions or feedback to help each of us live more emotionally self-regulated lives. Many of them will start off in the mental health space, though. The startup Mindstrong Health (founded by the former director of the National Institute of Mental Health) is examining how patients use their phones to provide clues about their mental health. The app doesn't actually see what the user is doing, but it tracks how quickly they tap, swipe, and scroll, detecting changes in behavior that could indicate a change in mental functioning. Similarly, NeuraMetrix, a San Francisco–based startup founded by two former cybersecurity executives, uses typing cadence as a biomarker for changes associated with neurological and mental functioning changes. Heart rate variability, something that many fitness wearables track, has known ties to emotional regulation,[16] so I expect emotional regulation tests will be available on smartwatches soon.

You'll want to get a numerical baseline of where you feel you are in your emotional regulation now. That way, you'll have a starting place to compare your progress against once you start self-experimenting. For your own self-assessment, here is a set of questions you can ask yourself now.

Self-Assessment of Emotional Regulation[17]

This survey examines certain core aspects of emotional regulation: the ability to monitor, evaluate, and modify emotions. It also includes self-soothing skills, impulse-control skills, self-awareness, and attentional control.

Remember, this is for you; you are not competing with anyone else. These questions provide a way for *you* to gauge where *you* think you're at on your emotional regulation for the time frame specified. Just answer as honestly as you can for your chosen time frame.

PICK YOUR TIME FRAME OF REFLECTION

For this survey, please answer all questions with a specific time frame in mind. To enable cross-comparisons, pick the same time frame for all self-assessments (e.g., if you picked "Last 30 days" for your self-assessment of executive function, pick 30 days here, too).

If you want to use your answers to this survey for assessing the effects of specific interventions during your daily self-experiments, choose "Last 24 hours." Or reflect every 24 hours if you want a more accurate baseline.

If you want to use it as a way to record your more general sense of your emotional regulation over a longer time period, choose "Last 30 days" or "Last 3 months."

LAST 24 HOURS ___ LAST 30 DAYS ___ LAST 3 MONTHS ___

Please indicate how often the following statements apply to you by writing the appropriate number from the scale below on the line beside each item. Your score for each question is the number you gave for that question. For example, If your answer to question 1 was 1, your score for that question is 1.

1 practically never: 0–10% of the time
2 sometimes: 11–35% of the time
3 about half the time: 36–65% of the time
4 most of the time: 66–90% of the time
5 almost always: 91–100% of the time

1. I noticed how I was feeling, including when I was upset.	
2. I could tell what my feelings were.	
3. I could change how I was feeling if I wanted to.	
4. I could soothe myself when I was upset.	
5. I could control my impulses by starting or stopping actions or words that were triggered automatically by strong feelings when I was upset.	
6. I could choose to think about something else when I was upset	

Score

Sub-Scores:

It's easy to focus on the overall score, but often the sub-scores of a quiz like this are actually the most useful. As you run self-experiments, you may discover that watching your changes on one or two of these questions provides the most useful measurement to you. Separately tracking the answers to this subset of questions going forward may provide specific value.

Overall Scoring Range: 18–90
Your final score is the sum of your answers to all the questions multiplied by 3. Higher scores indicate that you feel more confident in your abilities at regulating your emotions. Lower scores indicate that emotional regulation may be a bottleneck for you. If this is the case, it could be worth experimenting with interventions that target it.

Bottom third: 18–41
Middle: 42–66
Top third: 67–90

Mood Assessment

Because most assessments for emotional regulation are intended for long-term reflection (to capture your average levels over time), not for acute, in-the-moment assessments, you will use a mood assessment for your daily experiments. Use it in the following way: First, subject yourself to an emotional trigger (that is, think about something upsetting that causes you worry, anger, or fear), then take the Momentary Feelings Assessment (MFA) below. Then, try one of the interventions in this book to help you gain control over your emotions. After the intervention, take the MFA again. Calculate your new score to see if your mood has changed.

MFA: Momentary Feelings Assessment[18]	
Half of the following 18 feelings are pleasant and the other half are unpleasant. Indicate how much you agree that you are *currently* experiencing each feeling on a scale from 1 to 5, where 1 means very slightly or not at all, 3 means a medium amount, and 5 means extremely.	
I currently feel:	
1. Sad	
2. Motivated	
3. Angry	
4. Compassionate	

5. Frightened	
6. Grateful	
7. Disgusted	
8. Intrigued	
9. Anxious	
10. Purposeful	
11. Agitated	
12. Full of a sense of awe	
13. Regretful	
14. Energetic	
15. Testy	
16. Awake	
17. Unsettled	
18. Satisfied	

Scoring

Add up the scores for all odd-numbered feelings; that will be your "unpleasant" feelings score. This can range from 9 to 45. Add up the scores for all even-numbered feelings; that will be your "pleasant" feelings score. This can range from 9 to 45, too.

These sub-scores—the "pleasant" feelings score and the "unpleasant" feelings score—may be useful to track separately from the overall score. As you run self-experiments, you may discover that watching your changes on one or two of these questions provides the most useful measurement to you. Separately tracking the answers to this subset of questions going forward may provide specific value.

For your overall score, subtract your "unpleasant" feelings score from your "pleasant" feelings score.

Bottom third: −36 to −12
Middle: −13 to 11
Top third: 12 to 36

Lower overall scores indicate that your general mood is lower, higher scores that your mood is higher.

Now that we've looked at executive function and emotional self-regulation, let's turn our attention to an area you may have been waiting for: learning and memory.

TAKEAWAYS

1. Emotional self-regulation is being able to monitor, evaluate, and modify your emotions.
2. There are four steps to effectively manage an emotionally charged situation: (1) realize you are in an emotionally charged situation, (2) choose what to pay attention to within the situation, (3) choose how you interpret the situation, and (4) choose your response to the situation.
3. Multiple parts of the brain and body are active during emotional self-regulation: emotional (limbic) systems and frontal control regions (in the brain); the adrenal cortex (in the kidneys); and the peripheral nervous system, including fingertip temperature, pupil dilation (in the eyes), and skin (chemicals released).
4. You can measure your emotional self-regulation with surveys, but there may be more physiologically based ways to measure it in the future.

Chapter 9

Memory and Learning

"Memory is a complicated thing, a relative to truth,
but not its twin."
—Barbara Kingsolver

Time Investment: 15 minutes
Goal: To learn how memory works and
test your own

"Is this seat taken?"

I looked up and saw a smiling, wavy-haired boy. It was my first year at MIT, and I'd nabbed a seat near the front of the class in introductory psychology. "Go for it." I smiled back at the boy, whose name I would later learn was Nima Veiseh, and asked what made him interested in taking the class.

He chuckled. "My memory is kind of crazy and I'm hoping they can help me figure it out."

When our psychology professor flashed increasingly large sets of items on a screen and asked us to recall as many as we could after a short delay, a student a few seats down could recall nine items. Nima and I could remember only six. "That was a test of working memory," our professor explained.

"My memory difference must be some other type," Nima muttered to himself.

What Nima discovered much later, when he signed up to be in a UC-Irvine study, was that he was one of a few dozen people in the world who have been found to have a condition called by a few names: hyperthymestic syndrome, hyperthymesia, or highly superior autobiographical memory

(HSAM) — basically an inability to forget their experience.[1] Where I might remember the precise visual and verbal details of a conversation or a TV show for a few minutes to an hour afterward, Nima and his fellow hyperthymestics would remember it indefinitely. A hyperthymestic tends to answer correctly any verifiable question about their personal experiences 97 percent of the time, even if the experience occurred long ago.[2] Nima later used this ability to amass a detailed mental library of more than 1,500 paintings, which served as inspiration for his own unique art.[3]

Never forgetting sounded amazing to me. Nima confided that it is... but it also isn't. As much as we fear forgetting, it is actually a very useful mental cleaning process. The two minutes you spent in traffic the other day? Deleted. The two minutes you spent watching the Kentucky Derby with your grandfather the day before he died? Saved. For someone with hyperthymesia, though, both two-minute experiences may get remembered identically, even though their personal significance is radically different. One wonder of the human brain is that it runs on so little power — only around 20 watts. Yet, it may accomplish more than supercomputers that require a thousand times more power.[4] Our brain may accomplish this, at least in part, by forgetting.

Although Nima and others with hyperthymesia live their lives as if a film crew with inexhaustible film is rolling at all times, they may do so at a cost. Some hyperthymestics complained to researchers that their autobiographical memories were constantly being triggered, distracting them from organizing new information and making it difficult to retrieve and make use of the memories they already had.

Apart from their extraordinary autobiographical memory, other aspects of hyperthymestics' memories have been found to be average or even below average. Like the rest of us, they are susceptible to making false memories, too.[5] This makes the hyperthymestic brain seem a bit more attainable. So, what is the secret behind their tremendous autobiographical memory skills?

A 2015 study of hyperthymestics found that the more consistent the recall, the more likely the person was to have obsessive-compulsive tendencies — that is, to ruminate about events and replay them in their minds. The researchers believed that people like Nima encoded information in the usual way but *consolidated* their autobiographical memories differently.[6] Memory consolidation is a time-dependent process in the brain in which a recent, learned experience goes into long-term memory.[7] Perhaps hyperthymestics' secret sauce is that they consolidate better—by rehearsing and replaying events in their heads—than the rest of us do.

The Science Behind Memory and Learning

There are three stages to a memory (all types of memory, not just autobiographical): encoding, storage, and retrieval.

- **Encoding.** This begins when one of your senses detects something.
- **Storage.** That information gets filtered into a sensory memory that lasts just a few seconds. Then, a subset of that sensory memory gets filtered into short-term memory. Short-term memory lasts anywhere from a few seconds to many minutes, maybe even half an hour. Finally, a subset of that short-term memory gets turned into long-term memory, which can last a lifetime. There are a lot of obstacles to information passing through all of those stages and being encoded correctly, though. For instance, there can be interference from other information coming into your brain at the same time, or the information might not carry enough emotional significance to really make an impression. Many of these processes occur outside of your control, leading to some information you'd rather not remember being stored perfectly and other information that turns out to be critical being forgotten.
- **Retrieval.** The final stage is the one we typically associate with the act of remembering. It's the thing that's not happening successfully when we say, "I can *almost* remember! It's on the tip of my tongue."

Typically, we remember an event or a fact if it fulfills certain criteria: if it was emotionally intense, if it was surprising, if it fits in a pattern of things we've seen before (or, sometimes, if we simply try to remember it).[8] If it doesn't fulfill any of those criteria, we tend to forget it. Remember the neuroscientists' rhymes we discussed in Part I? "Neurons that fire together wire together" and "neurons that fall out of sync lose their link." The second one is, at a neuronal level, about the process of forgetting. Much as two friends will drift apart if they used to have a lot in common but now find less to say to each other, if the neurons stop firing together, they tend to fall out of touch. Falling out of touch is the brain's way of forgetting.

Where Do Memory and Learning
Happen in Your Brain?

There are four main types of memory. Each memory type is useful for different aspects of life functioning. We've just discussed autobiographical memory, which involves multiple areas, including the amygdala. In "The New IQ" chapter, we discussed working memory, the ability to hold more information in mind long enough to manipulate it—for a few seconds to a minute. In brain scans, the regions involved in working memory are mostly constrained to the front of the brain, whereas those involved in autobiographical memory also include the temporal lobes (located underneath your temples) and occipital lobes (near the back of your head).[9]

In school, we often rely on a type of memory called *semantic memory*. This is your memory for ideas, concepts, and facts. These include facts such as state capitals, vocabulary words in a foreign language class, or the elements of the periodic table. There's even a type of semantic memory that relates to the future instead of the past. This is *proscriptive memory,* or remembering to remember something. For instance, when you say to yourself, "Don't forget to pick up milk on the way home" you are invoking proscriptive memory. The hippocampus plays an important role in both episodic memory (of which autobiographical memory is a type) and semantic memory. In remembering, the front of the brain is still involved, but the temporal lobe may play a more important role.

When performing movements that require skill, we rely on another type of memory: *procedural memory.* You use this when you ride a bike or drive a car. With this type of memory, you had to pay attention at first, but soon it became automatic. The striatum area of the brain seems to be where your habits and procedural memories get stored and processed, and the front of the brain plays an absolutely critical role in getting them there. Encoding, the first stage of learning, relies heavily on working memory and therefore the prefrontal cortex—but when you know something well, you don't use the prefrontal cortex anymore. As we discussed in Part I, we can see that disparity in a scanner when we compare the activity of someone who is a novice at a task to someone who is expert at it. The novice's prefrontal cortex is afire with activity as they struggle to engage their working memory and other executive functions in order to do the task. In contrast, the expert's brain looks far quieter: they solve the problem simply by calling up automatic processes or long-term memories — this can involve various locations, including the striatum and the cerebellum.

This applies not only to tasks but also to learning attitudes or behaviors. As Adele Diamond of the University of British Columbia, a pioneering researcher in executive function, puts it: "A child may know intellectually (at the level of prefrontal cortex) that she should not hit another, but in the heat of the moment if that knowledge has not become automatic...the child will hit another. The only way something becomes automatic (becomes passed off from prefrontal cortex) is through action, repeated action. Nothing else will do."[10]

Key Regions for Learning and Memory: The Amygdala, Striatum, and Hippocampus[11]

FRONTAL CORTEX
STRIATUM
HIPPOCAMPUS
CEREBELLUM
AMYGDALA

Why Learning and Memory Are Important

Why does having high-functioning learning and memory help? For one, it saves you time. You do not have to check information; you just remember it. It's the difference between typing words into Google Translate and having a fluent conversation in a foreign language.

It also inspires confidence. Imagine a surgeon having to watch a You-Tube video every few minutes to brush up on technique. Imagine a TED speaker sorting through their notes as they gave their talk. Imagine a professional basketball player stopping midway down the court to check how to dribble.

Being able to learn quickly is increasingly important. More and more jobs expect a bachelor's degree or higher.[12] Artificial intelligence is automating away many low- and middle-skilled jobs; by 2030, between 75 and 375 million people globally are expected to change jobs; being able to pick up new skills quickly will help you adapt and capitalize on opportunities.[13] Even for those whose jobs are not threatened directly, the rate of new information that we are expected to keep on top of each day is overwhelmingly higher than it has ever been.[14] Being a fast learner with a good memory will help you stay cool amid all this change.

Testing Your Own Learning and Memory

If you want to find out if you have hyperthymesia like my friend Nima, you can go to UC-Irvine's study page and take their screening test.[15] For other types of learning and memory, you'll want to both get a self-reported baseline (where you feel you are in each of these areas) and performance-based baseline. That way, you'll have a starting place to compare your progress against once you start self-experimenting. You'll find performance-based tests on the web. For your own self-reported assessment, here is a set of questions you can ask yourself.

Self-Assessment of Memory and Learning[16]

The goal of this test is to provide a baseline for your current estimate of your learning and memory abilities. We're going to focus on two types of memory: episodic (personal experiences tied to places and times and emotions) and semantic (words, concepts, numbers).

Remember, you are not competing with anyone else. These questions provide a way for *you* to gauge where *you* think you're at on your learning and memory for the time frame specified. Just answer as honestly as you can for this specific time frame.

PICK YOUR TIME FRAME OF REFLECTION

For this survey, please answer all questions with a specific time frame in mind.

If you want to use your answers to this survey as a baseline or for assessing the effects of specific interventions during your daily self-experiments, choose "Last 24 hours."

If you want to use it as a way to record your more general sense of your memory and learning over a longer time period, choose "Last 30 days" or "Last 3 months."

LAST 24 HOURS ___ LAST 30 DAYS ___ LAST 3 MONTHS ___

Your score for each question is the number you gave for that question. For example, if your answer to question 1 was 1, your score is 1.

1	almost always: 91–100% of the time
2	most of the time: 66–90% of the time
3	about half the time: 36–65% of the time
4	sometimes: 11–35% of the time
5	practically never: 0–10% of the time
n/a	It didn't come up.

I struggled to learn new numbers.	
I struggled to remember numbers I already knew.	
I struggled to learn new facts.	
I struggled to remember facts I already knew.	
I struggled to learn new words.	
I struggled to remember words that I used to know.	
I struggled to learn names and/or faces.	
I struggled to recall names and/or faces I already knew.	
I struggled to remember where I put things (keys, wallet, car).	
I struggled to remember what I was about to do.	
I accidentally double-booked myself or promised the same thing to two people because I forgot my commitments.	
I forgot to do routine tasks (brush my hair, take my medicine, pay my bills).	
I struggled to learn from reading material.	
I struggled to remember things I had read.	
I struggled to remember things I had heard.	
I struggled to remember what I was saying, doing, and/or thinking if I got interrupted in the middle.	

I struggled to learn and/or remember directions and/or instructions that I had just learned.	
I struggled to remember directions and/or instructions that I have known for a long time.	

Score

Sub-Scores

It's easy to focus on the overall score, but often the sub-scores of a quiz like this are actually the most useful. As you run self-experiments, you may discover that watching your changes on a subset of these questions provides the most useful measurement to you. Pick out 3 to 5 questions from the Self-Assessment of Memory and Learning that most resonated with you — the ones that most match your internal definitions of memory and learning. Separately tracking the answers to this subset of questions going forward may provide specific value.

Overall Scoring Range: 18–90

If you answered "n/a," don't include that question in your final score.

Your final score will be the sum of your answers to all the questions. Divide that sum by a second number. This number should be the product of the number 5 multiplied by the total number of questions with answers that were *not* "n/a."

Now, take the result of the division and multiply it by 72.

What tier is your score in?

Bottom third score: 18–41
Middle score: 42–66
Top third score: 67–90

Lower scores indicate that learning and/or memory may be a bottleneck for you. If this is the case, it could be worth experimenting with interventions that target it. If your score is getting higher over time, that's a positive indication that your neurohacking experiments may be working.

Performance-Based Tests of Memory and Learning

If you want a quick way of testing your word-learning abilities, try the following test:

Word-Learning Test

Materials

- Timer
- Word bank #1 and #2 (included below)
- Blank page to record all words you remember
- Something to write with

Directions

1. Set your timer for 1 minute.
2. Press start and look at the 20 words in word bank #1. Commit as many to memory as you can. Then, cover up the word bank.
3. Take a break of 1 minute—don't think about the words!
4. Then, give yourself another minute to write out as many words as you can remember.
5. Uncover the word bank. What percentage of the 20 words did you remember?

Word bank #1 (for a pre-intervention test)

Dog
Table
Mountain
Hat
Train
Democracy
Tulip
Quinoa
Ripple
Echo

Wise
Disk
Fireplace
Wonder
Disapproval
Maze
Woods
Knee
Painting
Eyes

Word bank #2 (for a post-intervention test)

Cat
Couch
Sky
Coat
Plane
Monarchy
Daisy
Turmeric
Hole
Silence
Foolish
Rectangle
Microwave
Discouragement
Celebration
Game
Desert
Earlobe
Sculpture
Hands

Next, we'll turn our attention to the last of our potential mental targets: creativity.

TAKEAWAYS

1. Memory comes in short-term and long-term varieties. Episodic memory is about the stories of your life, the episodes, whereas semantic memory is about concepts and facts, and procedural memory is about how to do things.
2. Learning happens in two stages—encoding and storage—and recalling what you've learned involves retrieval.
3. Becoming a faster learner will help you adapt and stay on top of your career, despite the threat of global workforce changes occurring due to automation. Improving your memory will help you enjoy and make use of what you already know.

Chapter 10

Creativity

"You can't use up creativity. The more you use, the more you have."
— Maya Angelou

Time Investment: 17 minutes
Goal: To better understand creativity and learn to test your own

During the winter of 2015, I was invited to give a talk at an education conference in Suzhou, China. While I was there, the conference organizers invited the other international speakers and me on a grand tour of innovative K–8 schools in the Shanghai area. While I encountered many new ideas—for example, middle school teachers routinely publish research, just as university professors do in the US!—one classroom stood out to me. Like most of the schools we visited, this school prided itself on its academics. But as my colleagues and I filtered into the art room, we were treated to an entirely different facet of the educational experience. Every wall, every flat surface, every corner, every shelf was crammed with paintings, drawings, murals, and sculptures. There were a surprising number of pieces that possessed an emotional expressiveness and controlled creativity that even a professional adult artist would have been proud to call their own.

The range in quality was all the more startling; there were masterpieces stacked next to works that looked little better than stick drawings. When the art teacher saw my confused expression, he laughed. This semester, he said through our translator, the students were being graded as much on the quantity of what they produced as the quality. Their ultimate grade was not just dependent on *how good* their artworks were; it was also about *how many* they could produce. At this point in the semester, the room was almost full of all that creative productivity.

Something about this crowded art room tickled my memory. As I was leaving the room, I suddenly remembered my own seventh-grade English teacher. She had given my classmates and me an assignment to create a writing journal. Working on this was part of our final grade, but we would be graded not on the content of the journal, only on the quantity of pages that we produced and our conscientiousness in doing so. All of us were astonished and a bit suspicious. "What's the catch?" one of my classmates asked, but the teacher smiled and repeated herself. "No catch. Write about whatever you want, however you want. Diary entries, poetry, song lyrics, stories, essays, shopping lists...just write."

Both the Chinese art teacher and my seventh-grade English teacher, as it turns out, were onto something. You'll see what I mean soon.

The Science of Creativity

What is creativity? Who really knows?! Creativity is a devilishly difficult thing to define—let alone measure. In fact, I agonized for quite a while before caving and including it in this book. Since our goal as neurohackers is to improve mental abilities with an impact on daily life, however, we'll look at creativity. But we'll use a fairly narrow, quantifiable definition. My hope is that this will be of more real-world use. We know creativity has to have something to do with novelty. It's about creating, so it's got to be about making things. Also, the thing you make must be useful or valuable. So, here's the definition I've heard many researchers go by, at least informally: creativity is about the number of useful and novel things you produce in a given time period.

Practically speaking, that would mean that someone who is more creative would be able to produce a higher *quantity* of novel, useful items in a given time frame than someone less creative.

Some people believe that genius levels of IQ are required for creativity, but multiple studies have challenged this notion. Since IQ tests for adults are designed to be harder than those designed for children, an average-scoring child would score lower than an average-scoring adult on an adult IQ test. Yet, when teams were tasked with building the tallest structure they could out of just a ball of string, a roll of tape, a stack of spaghetti, and a marshmallow, teams of kindergartners outperformed teams of adults — including teams of lawyers and business school students. The kindergartners dominated.[1]

Multiple researchers have found data supporting what they call the "threshold hypothesis."[2] From looking at individuals' performances, they discovered that having a somewhat above-average IQ—say 120 (which is above but not drastically above average) — is what's necessary for extreme creativity. To put that 120 in context, most people score between 80 and 120 on IQ tests, with the average being 100. Scoring 130 or higher could, depending on the specific type of IQ test, put you in the top 2.2 percent of IQ test takers. Decades ago, IQ test makers used to call such a score a "genius."[3]

However, a 2013 German study puts the "threshold hypothesis" in a new perspective.[4] While previous studies tended to use small groups of college students as their subjects, these researchers looked at a sample of nearly 300 people, and they included people of more wide-ranging educational backgrounds and varying IQ levels. First, they gave them classic psychological creativity tests. For example, one type of creativity test might be: "Come up with as many uses for a brick as you can in one minute." Another creativity test might be: "To the nearest cubic inch, how much soil is there in a 3 foot by 2 foot by 2 foot hole?" (Here, the answer is zero, because you're supposed to recall that it was a hole.)[5] They also asked them about their real-world creative achievements. (Have you ever designed or sewed clothing? How many times? Have you ever sold any?) From all this, the researchers discovered something surprising.

The threshold they found for performance on the classic creativity tests was actually much lower than had previously been found. For some, it was way below 120—it was actually 85. When it came to creative achievement in the real world, they found that personality and IQ both played larger roles than previous studies had shown. In high-IQ individuals, being "open to new experiences" (a core personality trait) was predictive of creativity. Furthermore, when it came to creative achievement, they didn't see the diminishing gains of IQ that previous studies had found. They found that higher-IQ individuals actually had more creative *achievements* than lower-IQ individuals, all the way up to the highest-IQ individuals in their study.

Their results suggest that people of lower IQs may be more capable of creative achievement than they are getting a chance to express, whereas higher-IQ individuals seem to more often get opportunities to explore and get credit for their creative impulses. Privilege, rather than potential, may be gating creative expression in our society.

Beyond IQ, researchers have found that certain types of thinking tend to give rise to new ideas, inventions, or approaches. While *linear thinking* is associated with logic and reasoning, creativity is more associated with *divergent thinking* (generating creative ideas by coming up with many possible solutions), *lateral thinking* (solving problems through an indirect approach — often by seeing things from a different perspective or in a different light), and *convergent* or *synthetical thinking*. There's also fluency of thinking — how fast you think of ideas and how many ideas you come up with, but not necessarily whether any of the ideas are particularly good.

What Boosts Creativity?

"No one in his or her right mind would argue that quantity guarantees quality," bestselling novelist Stephen King writes, "but to suggest that quantity never produces quality strikes me as snobbish, inane and demonstrably untrue."[6]

Many of the people we think of as the most creative in history had staggeringly prolific levels of output in their careers. Charles Darwin published around 120 scholarly papers, Albert Einstein published around 250 papers, and Sigmund Freud published 330 papers. Thomas Edison held nearly 2,000 patents on his inventions. Johann Sebastian Bach composed more than 1,000 musical pieces, and Pablo Picasso is credited with more than 20,000 paintings, sculptures, and drawings.[7]

Going back to the journal writing challenge set by my own seventh-grade teacher, my classmates all commented that it helped them become less self-conscious. Furthermore, even those who had previously hated writing started to enjoy it. For me, it acted like a creativity release valve. Perpetually dogged by perfectionism, I found that the new stakes finally loosened me up.

Without the fear of judgment, I spontaneously produced poems, songs, and short stories. Boosting the quantity of your output could help you become more creative, as we discussed above. A positive, elevated, and open mood also seems to disinhibit and allow for more ideas to come.[8] In this state, you are unhindered by internal criticism.

Lest you think that creativity is all about being in a good mood and just waiting for magic, it's clear that the emotional recipe for creativity is, well, not fully clear. Emotional intensity and even ambivalence seem to play a role in creativity, too.[9] Researchers used to believe that positive moods led to creativity, but recent research has revealed a messier truth. High-intensity feelings, even if they are negative, can lead to completing a set goal, whereas low-intensity feelings, again, even if they are negative, enable us to think more broadly, more diffusely—the kind of thinking necessary to shift perspective and "see the big picture."[10] Expertise matters, too. Recall that two of the criteria for creativity are novelty and usefulness. If you know nothing about a field, you might get lucky and produce a few creative ideas (especially if you are expert in another area and are transferring your skills from there), but how can you know what is useful or novel in this new area without expertise? Being at the right challenge for your skill level is key.[11]

In short, the idea is practicing a lot, being in the right mood—and more the magnitude than whether you're happy or sad—and don't forget you need to have a level of expertise in that area, of course. All of these will affect your creativity.

Where Creativity Lives in the Brain

Ever heard that the right brain is creative and the left brain is logical? Sorry to be the bearer of bad news, but recent brain imaging data has debunked that idea. The old idea seems to have come from language centers being in the left hemisphere and spatial abilities being in the right, but recent findings show a far more complex and interesting picture.[12]

Recent studies of people lying in brain scanners and performing tasks requiring creativity or creative thinking showed that they tended to engage not just one part (say, the right or left hemisphere of the brain) but multiple large-scale brain networks that run through the frontal area, the temporal lobes, and the limbic system.[13] This could be seen not just in randomly selected people, but also in people who are indisputably creative, such as free-style rappers and jazz improvisers.[14]

Why Increase Your Creativity?

There are many reasons why interventions designed to boost your creativity might be worth pursuing. First of all, creativity brings a special kind of focused joy: flow. Flow, also known as being "in the zone" or being immersed, fully absorbed in a feeling of energized focus, is an inherently pleasurable experience. It's common when in the throes of creativity.[15] Many artists, scientists, and performers report that they lose track of time and lose any sense of themselves when they are immersed in their craft.

Second, creativity may lead to interesting careers. In the 21st century, we are going to need a lot of innovative solutions for everything from climate change to how we should try to coexist with AI. Not to mention how to bridge the widening gap between rich and poor, how to make our food supply chains more sustainable, or how to travel to other planets. If you want to be part of solving these problems—some of which may involve lucrative career paths—you may want to boost your creativity.

Creativity also offers the possibility of personal glory and immortality. If you want fame or fortune now, doing the same things as everyone else but faster and more reliably could be your ticket. If you want to be remembered forever, however, creativity is the path to get there. Maybe when you unlock more of your creativity you will become the next Coco Chanel, Edison, Marie Curie, Beethoven, or Dostoyevsky. Your art could be in museums, your music could be played on everyone's devices, your company could be publicly traded, your cure for a disease could save millions, your inventions could be sold in stores everywhere.

Testing Your Own Creativity

You'll want to get a self-reported baseline (where you feel you are in each of these areas) and establish a performance-based baseline. That way, you'll have a starting place to compare your progress against, once you start self-experimenting. For your own self-reported assessment, here is a set of questions you can ask yourself. You'll find performance-based tests on the web.

Self-Assessment of Creativity[16]

This survey examines certain core aspects of creativity: usefulness, novelty, lateral thinking, divergent thinking, ability to go into flow, quantity and quality of work, creativity across more than one domain, using "hard skills" (acquired skills) to be creative, and so on.

Remember, you are not competing with anyone else. These questions provide a way for *you* to gauge where *you* think you're at on your creativity for the time frame specified. Just answer as honestly as you can for this specific time frame.

PICK YOUR TIME FRAME OF REFLECTION

For this survey, please answer all questions with a specific time frame in mind:

If you want to use your answers to this survey as a baseline or for assessing the effects of specific interventions during your daily self-experiments, choose "Last 24 hours."

If you want to use it as a way to record your more general sense of your creativity over a longer time period, choose "Last 30 days" or "Last 3 months."

LAST 24 HOURS __ LAST 30 DAYS __ LAST 3 MONTHS __

Answer the questions below about an area or areas where you have expressed creativity. Feel free to choose from the creative categories below or come up with your own:

Arts: painting, drawing, sculpture, music, dance, writing, culinary arts, DIY and crafts

Design: digital, clothing, architecture, interior, landscape

Entertainment and persuasion: animation, humor/comedy, drama/theater, marketing and advertising

Caring: healthcare diagnoses, teaching, interpersonal conflict resolution

Organizational leadership or founding: political or economic policy

Scientific, mathematical, or engineering: discoveries and/or inventions

For each statement below, answer 1–5 or n/a:
1 I disagree strongly.
2 I disagree.
3 I am neutral.
4 I agree.
5 I agree strongly.
n/a It didn't come up.

The quantity of my creative work was high; I was prolific.	
My creative work was novel and/or useful.	
I regularly achieved "flow" (a state of being fully immersed in an activity, having an energized focus, and enjoying the process — not just the outcome — of the activity).	
I was good at lateral thinking (solving problems through an indirect approach, often by seeing things from a different perspective).	
I was good at divergent thinking (generating creative ideas by coming up with many possible solutions).	
In my creative activity, I used hard skills (learned abilities and enhanced through practice).	
I was creative in at least one area.	
I was creative across multiple domains.	
I designed a new program or course.	
I launched a new initiative(s) that attracted attention and a following.	
I successfully passed a new rule or regulation.	
I resolved a conflict previously thought to be intractable.	
I solved a problem that was deemed unsolvable or did something that has never been done before.	
My creative work was praised by people who are skilled and/or experienced in this area.	
I won awards or competitions or was selected for competitive groups, at a local level or beyond (in my school, company, my town, etc.).	
My work was covered locally or beyond.	
My work was published, produced, manufactured, performed, or shared with strangers in some way.	
People paid money for my creative works.	

Scoring

Sub-Scores:

It's easy to focus on the overall score, but often the sub-scores are actually the most useful. As you run self-experiments, you may discover that watching your changes on a subset of these questions provides the most useful measurement to you. Pick out 3 to 5 questions from the Self-Assessment of Creativity that most resonated with you — the ones that most match your internal definition of creativity. Separately tracking the answers to this subset of questions going forward may provide specific value to you.

Overall Scoring Range: 18–90
Your score for each question is the number you gave for that question. For example, if your answer to question 1 was 1, your score is 1. If you answered "n/a," don't include that question in your final score. Your final score will be the sum of your answers to all the questions. Divide that sum by a second number. This number should be the product of the number 5 multiplied by the total number of questions with answers that were *not* "n/a." Now, take the result of the division and multiply it by 72. What tier is your score in?

Bottom third score: 18–41
Middle score: 42–66
Top third score: 67–90

Lower scores indicate that creativity may be a bottleneck for you. If this is the case, it could be worth experimenting with interventions that target it. If your score is getting higher over time, that's a positive indication that your neurohacking experiments may be working.

Creative Output Tracking Assessment
The following self-assessment is less of a test and more of a way of tracking your creativity during a specific time period — say, during a self-experiment. You can use it instead of the self-assessment or in addition.
Pick your time frame of reflection (e.g., 30 days, 3 months)
What area of creativity will you assess (choose one of the fields above)?
How much time did you spend on your creative activity (inventing, painting, etc.)?

How much did you produce (number of paintings, poems, etc.)?
How did you *feel* about your creative activities (how focused were you, how often did you go into flow)?
What feedback, if any, did you receive on your work (mentor praise, media mentions, citations in your field, etc.)?

Performance-Based Tests of Creativity

Alternative-Uses Test

Materials

- Timer
- Prompts below
- Something to write with

Directions

Come up with as many uses for one commonplace object as you can in a minute!

Example: Rock—a paperweight, a weapon, a doorstop, a worry stone, a jewelry ornament, ingredient for driveway gravel, ingredient for fish-tank gravel, element in a rock garden, material for a sculpture, irritant in your shoe. (10 items in a minute)

Prompts

- Basket
- Disk
- Desk
- Map

- Cabinet
- Cigarette
- Shirt
- Camera
- Newspaper

Scoring

Quality and quantity both count. Since you'll be judging your own work, try to be as fair as possible. For instance, if your prompt is "basket" and you answer "carry berries," that's not as creative as "hat" or "fodder for a fire." Coming up with an unconventional use earns more creativity points. To earn the most points, you'll need to be both fast and creative.

Verbal-Fluency Test

Materials

- Random letter generator: https://www.randomlettergenerator.com/
- Timer
- Prompts below
- Paper
- Something to write with

Directions

Use the random letter generator to generate a letter, then come up with as many words that start with that letter as you can in 1 minute.

Example: I got the letter "R" and generated 19 words in 1 minute:

Ribald

Ribbing

Ribs

Ribbed

Ribber

Ripped

Ripping

Rips

Rip

Ripest

Rotten

Rot

Rots

Rottenest

Rotted

Rock

Rocks

Rocking

Rocker

Now that you've learned a bit about these four potential mental targets, you may know which one you want to focus on first. But if you haven't decided yet, don't worry—the next chapter will help you narrow down a first target.

TAKEAWAYS

1. Creativity is the ability to come up with new, useful ideas and creations.
2. In order to get high-quality creativity, it helps to focus on quantity—generating more and more products. Also, it helps to have a skill or an area of expertise to tap into—or multiple areas, so that you can find unusual connections between them.
3. Being in a positive mood may aid creativity, but emotional intensity—regardless of whether it's positive or negative—may play an even deeper role.
4. People who score lower on IQ tests tend to be given fewer opportunities to exercise their creativity, but lower IQ does not seem to be an intrinsic impediment to creativity. Among higher-IQ individuals,

those who were more open to new experiences were found to be more creative.

5. You can measure your creativity over time in many ways, including using psychological tests — such as the Alternative-Uses Test or the Creative Output Tracking Assessment to count the number of creative acts you achieve over a set time (number of poems composed per month, number of articles written per year, etc.) — or using functional tests.

Chapter 11

Choosing Your Mental Target

Time Investment: 9 minutes

Goal: To choose the mental ability you want to target with your first neurohacking experiment

If, after reading about each of the four mental targets, you're still unsure of which mental ability to target, you may want to make your choice based on one of two things: jaggedness or wobble.

Imagine looking at a cozy, small town from a distance. Many of the houses are of similar heights. Perhaps there is a courthouse or a place of worship that is higher than the other buildings, but they are within a few stories of one another. Now, consider the skyline of New York City as it looks from across the Hudson River. The range of building heights—from skyscrapers to one-story gas stations that you might just barely be able to see—is vast.

Full NYC Skyline[1]

How would it look if we cut off all the tallest buildings? Doesn't look much like New York City anymore.

NYC Skyline Without the Tallest Buildings

How would it look if we removed all the shortest buildings? This doesn't look like New York City either.

NYC Skyline Without the Shortest Buildings

To understand your own brain, you need to see it in all of its uneven glory, its greatest strengths and its greatest weaknesses.

Jaggedness: Differences Among Your Mental Abilities

Every person's brain is made up of a mass of different skills and abilities, containing a mix of towering intellectual or emotional strengths alongside areas of relative weakness. Jaggedness is a term that researchers use to describe the abilities of a person who possesses both great strengths and great weaknesses compared to an average of the general population.[2] Numerically, jaggedness is the difference between the average and the extremes of a person's performance across different abilities (such as attention, memory, and creativity). For this reason, even two brains that are both considered of "average" IQ may have very different strengths and weaknesses. Knowing your level of jaggedness can clue you in to detecting a possible bottleneck, or an area of weakness in an ability that is required for accomplishing a task. Because you lack that one area needed to do a task, you never get a chance to exercise your strengths in other areas. For instance, if you were a terrific singer but couldn't read music, and reading music was a requirement for gaining acceptance into a choir, reading music would be your bottleneck. As neurohackers, we look for bottlenecks in *abilities* rather than in *skills*. The theory goes that if you detect a bottleneck and then successfully eliminate it, you'll finally gain access to previously blocked abilities.

In school and in the workplace, when someone has a significant bottleneck, they will often get labeled and sorted according to their weaknesses, rather than their strengths; the bottleneck prevents their strengths from being revealed. A heartbroken mother once confessed to me her daughter's conundrum: the girl's spoken verbal abilities revealed precocious abstract thought. Other aspects of her intelligence measured in the 99th percentile as well. Her slow processing speed was a bottleneck, however. As a result, she scored in the 10th percentile in reading. In some school environments, a child with this amount of jaggedness might have been called "twice exceptional" because she would have been labeled as both "gifted and talented" and "special needs."[3] Unfortunately, in this girl's school, because she could not keep up with the pace of reading expected in the mainstream classroom, her teachers classified her by her weaknesses rather than her strengths. School became hell for her. In addition to her work with a learning specialist, I suggested to the girl's mother that they conduct neurohacking experiments together that focused on upgrading the girl's bottleneck: her processing speed.

MEASURING JAGGEDNESS AND BOTTLENECKS

So how do you measure jaggedness and bottlenecks? Look at the lowest score, the highest score, and the average score across tested abilities. Take assessments for multiple mental areas and review your results. If the range of your scores is narrow, your profile is not very jagged, and it's unlikely you have a bottleneck. If the difference between your highest and lowest scoring areas is very large, your profile is jagged. In this case, you'll likely gain the most from upgrading one of your lowest-scoring areas. If all of your abilities are fairly strong, don't worry about bottlenecks. Instead, you may want to focus either on honing your highest-performing area or on making your performance more consistent.

Attention is a common bottleneck for other mental abilities, especially among children,[4] people who struggle with depression[5] or anxiety,[6] people who get insufficient sleep[7] or deal with brain fog,[8] and people with ADHD.[9] Without control over your attention, you may struggle to reliably learn, remember, be creative, or control your temper. While you may have tremendous untapped potential in your other domains of mental performance (learning and memory, creativity, emotional regulation, and so on), until you gain control over your attentional bottleneck, all of that mental potential will remain exactly that. Thus, if you score significantly lower in executive function than you do in the other three domains, I highly recommend you start your neurohacking journey by focusing on your executive function. You may find that improving it automatically improves the other areas — not because you've directly improved them, but because you've removed a bottleneck and thus unblocked them.

Wobble: Differences in Your Performance over Time

While jaggedness describes the differences in strengths and weaknesses across different mental abilities, wobble describes the differences in strengths and weaknesses within one mental ability — but across time. Any mental ability with a lot of wobble could be a bottleneck area that is potentially worth targeting with neurohacking self-experiments.

Numerically, wobble is the difference between the average and the extremes of a person's performance in one domain (such as executive function) across time. A person can have a wide range within an ability depending on the time of day, their age, whether they are a novice or an expert (novices tend to have high wobble and experts tend to have less, allowing them to perform more consistently), and a variety of other factors, including overall health.[10] Upgrading areas of high wobble can be especially rewarding, because it can unlock potential. By looking at your scores over time, you'll be able to assess your wobble. If you notice that your scores within one mental ability vary more over time than other abilities, that mental ability is potentially a bottleneck for your overall performance.

There are two definitions of wobble that I like to use. One focuses on your peaks: it is the difference between your average mental-performance day and your best. The second approach is the difference between your average mental performance day and your worst. The first approach shows you the potential you could neurohack your way into, and the second approach reveals your biggest mental performance liability currently.

CHRONOTYPE-RELATED WOBBLE

While wobble can show up in specific mental abilities, I was surprised when I saw how much chronotype (roughly, whether you're a "morning person" or a "night owl") creates wobble in overall mental performance. In hindsight, it shouldn't have surprised me. Growing up, I witnessed both ends of the chronotype spectrum at my kitchen table. Breakfast was a study in contrasts, whenever my mother could convince all of us to wake up in time to eat with her. My father, sister, and I all huddled miserably at our respective corners of the table, sipping coffee, hiding yawns, and trying to smile and nod at the appropriate parts of my mother's giddy, high-octane speech. The three of us watched, with a mix of horror and awe, as she frolicked around the kitchen, some kind of magical "morning person" elixir running through only her veins. "How is she so awake right now?" I would whisper to my sister. "Are you sure we're related?" she would whisper back.

No matter your chronotype, your mental performance varies dramatically depending on the time of day.[11] The difference between top and bottom scores can be extreme for the same person throughout a single day. I remember taking practice tests for the SAT and noticing that I had a 50- to 100-point difference between scores depending on the time of day that I took the exam. For most of us, our chronotype is later—toward the "night owl" end of the spectrum—during adolescence. Most schools put their students at a disadvantage by beginning early in the morning. We now know that taking classes out of sync with our chronotype results in poorer performance.[12]

In a study published in 2018 conducted on 56 healthy participants in the United Kingdom, evening types and morning types were tested at several times of day on two different attention tasks. Chronotype was identified through self-report as well as through objective measures such as a movement tracker and samples of saliva (looking at cortisol and melatonin levels). The difference between the best and worst version of the same person's performance over a single day was dramatic: the morning type starts the day scoring nearly 100 on a test of executive function but, as the day progresses, the morning type's performance crashes. By their lowest point in the day (around 2:00 p.m.), the morning types are scoring closer to 90. So, in a measurable sense, they are almost 10 percent worse mentally than they were just hours before. Tasks they can handle easily in the morning can be quite a struggle by early afternoon.[13]

It's clear that wobble affects mental performance, but what can be done about it? In a perfect world, we would all follow schedules that complemented our chronotype.[14] For instance, night owls might never schedule meetings earlier than 11:00 a.m. For many of us, however, our daily schedules are set by other people. While I truly hope that schools and workplaces become more flexible, until then, we can optimize ourselves by neurohacking when we know we'll be at our worst. If you're a night owl, neurohacking first thing in the morning may give you a boost to overcome your natural tendencies. For instance, you could run a self-experiment comparing two interventions to see which gave you the most immediate mental performance boost right after breakfast. For morning types forced to take an evening exam, say, they could run their self-experiments in the late afternoon. Again, your best performances are likely to come when you work at times that suit your own chronotype, but if you have to work under adverse circumstances, pit two interventions against each other and use whichever intervention gives you the most immediate boost when you're at your worst.

AGE, EXPERIENCE, AND OTHER CAUSES OF WOBBLE

There are other sources of wobble. Some researchers have found that children and older adults both tend to have more wobble in their mental performance than people in the middle of their lives.[15] In addition, they have found that people who struggle with their health often struggle with high wobble in their mental performance, too.[16]

Another cause of wobble can be inexperience. When giving time estimates to managers, for instance, it's very common for inexperienced workers to complete work in highly variable time frames.[17] This kind of wobble tends to even out with experience; after someone has done a task many times, they know how long it will take them to perform it. As a manager, I often noticed that with inexperienced engineers, for instance, their time estimates for how long it would take them to do a task varied far more than they did for more experienced engineers. The more experienced engineers were not always more accurate at predicting how long something would take, but they were more reliable. For instance, I noticed that they might consistently say a task would take one and a half times as long as it actually ended up taking them. For the inexperienced engineers, however, their time estimates ended up being fairly random; they had "high wobble."

TAKEAWAYS

1. Consider targeting mental abilities with jaggedness or wobble when choosing where to focus your self-experiments; both can indicate a bottleneck somewhere in your mental performance. Targeting a mental area that is a bottleneck can lead to a more successful upgrade.
2. You can compare your results across the surveys in chapters 7 through 10 to assess your jaggedness.
3. If you take the performance-based tests in chapters 7 through 10 repeatedly (use the instructions in "The Nuts and Bolts" chapter for guidance), you will be able to detect wobble.

Chapter 12

Life Scoring

"If you don't know where you're going, you'll end up someplace else."
—Yogi Berra

Time Investment: 12 minutes
Goal: To learn how to track your productivity and satisfaction throughout life. These will be sanity checks to ensure your neurohacking has real-world significance.

What's the point of upgrading your mental abilities if you don't apply them to improving your life overall? In this chapter, we'll explore two measures: your Say to Do score and your Life Satisfaction score. These measures provide a way for you to periodically assess how your neurohacking relates to these less scientific but oh-so-meaningful aspects of life that make up the quality of your life.

Say to Do

To explore the ins and outs of New Year's resolutions, I started tracking my own annual Say to Do scores in 2011. Every year, I make New Year's resolutions that are concrete—they can't be something vague like, "Be a better person by next year." I set goals in my personal life and in my professional life. The goals have to be SMART—as in specific, measurable, actionable, relevant, and time-based.[1] Then, I begin tracking my Say to Do score—the percentage of things that I say I will do that I actually end up doing.

I break down that year's goals into milestones that I am supposed to hit each quarter. Within each quarter, I have phone check-ins with an accountability buddy about what I planned to accomplish during a specific time frame; we typically meet every week to make our promises for the next week and report on the previous week. We also have special meetings at the beginning of each year and each quarter to confirm our annual and quarterly goals with each other. While all of this may sound like a lot of meetings, I suspect that the precise frequency matters less than the regularity; I've heard of other accountability buddies who met less frequently but still enjoyed success.

Let's look at some examples of Say to Do in action. My top professional goal the year that this manuscript was due read as follows: by December 31, email the finished first draft of the book's manuscript to my editor. By the end of the first quarter (March 31), finish a very messy first draft (that only I would see), one that hit my ultimate word-count goal. By the end of the second quarter (June 30), email two chapters to my editor for her feedback on whether my style and structure are heading in a good direction. My top personal goal that same year was to plan my wedding. My goal by the end of the first quarter I started planning was to have picked a wedding date, reserved a venue, and sent out invitations.

For me, ultimately, the most flexible and powerful tool to keep track of my resolutions was the spreadsheet; I liked being able to make graphs automatically. For others, a paper journal might be the answer. Regardless, tracking made it possible to compare what I had planned on doing to what I actually accomplished. I could also see my successes, failures, and learning year over year. Although my system has changed and evolved over time, enough stayed the same that I was able to create graphs for a 2018 Quantified Self talk where I compared my old Say to Do scores from 2011 to my newest scores.[2]

In addition to my Say to Do scores, I tracked my Life Satisfaction score. I did this because I wanted to make sure I was committing to things that actually made me happier. If I performed well in my Say to Do scores but didn't feel satisfied, that would trigger concerns that I was either aiming at entirely the wrong things (things I didn't actually care about but thought I *should*) or aiming poorly (at goals that were too easy). Similarly, if I got very little done but felt pretty satisfied, that would also imply that I aimed poorly (set too many goals). Part of managing both my Say to Do and Life Satisfaction scores was learning not to overcommit.

A LOOK AT THE LIFE SATISFACTION SCORE

To track my Life Satisfaction score, I started with what I learned from an MIT class called Designing Your Life. The teacher was Lauren Handel Zander, the firebrand founder of the Handel Group, an executive coaching firm in New York City. She had taught her life design courses at Stanford Business School, life coached Fortune 500 executives, and was not afraid to challenge anyone to stop being dishonest with themselves and finally admit to what they wanted in life. (Oscar winner Forest Whitaker was quoted as saying, "Lauren Zander does not mince words."[3]) Her obsession was personal responsibility and integrity, and I loved it. My favorite exercise in her course—and the one I ended up adapting and using for years afterward—was a self-assessment called the 18 Areas. In it, Zander had identified 18 key areas of life on which to focus self-improvement (work, relationships, health, home, finances, spirituality, and so on).[4]

Zander made each person rate aspects of their lives on a scale from 1 to 10. The self-assessments forces you to evaluate where you are at that specific moment relative to where you wished you were. On Zander's scale, 1 was awful, 5 was "a 6 that's been around for a while," 6 was "weak but not painful," and 10 was "amazing but unsustainable"—essentially, a peak moment. So, 9 was really the number to aim for in an ongoing way. The goal was to write your dream version of your life in each of those areas, contrast that with your current version of your life, and then explain to yourself why you weren't living your dream yet. Learning how to challenge your own explanation for that disconnect was much of the work of the class.

When the class ended, I realized that I wanted to keep taking my own pulse on all the life areas. I began reading about positive psychology—a relatively new area in psychology that uses scientific methods to study why some people thrive and flourish. This fascinating field has created its own instruments and tools to evaluate happiness and well-being. I discovered a parallel idea to the 18 Areas in positive psychology called the Wheel of Life.[5] There was also a happiness survey that I tried to take at least once a month. It is built on positive psychology pioneer Martin Seligman's finding that psychological well-being and happiness are composed of five feelings: positive emotion, engagement, relationships, meaning, and accomplishment, or

"PERMA."[6] I evaluated myself in each of those five areas, asking myself how frequently I experience each of those feelings (from "almost never" to "most of the time").

I decided to create my own life tracking system. It shared some ideas not only with the Wheel of Life and the 18 Areas, but also with a popular survey called the Satisfaction with Life Scale, which was developed by the legendary Ed Diener, another of the fathers of positive psychology.[7]

By taking these surveys repeatedly over time, I began thinking of my current Life Satisfaction score as a less abstract, more controllable thing. Just as your weight and resting heart rate are due to a mix of genetics and lifestyle, I began to appreciate that while my life satisfaction was partly inherited, there was a big part that I could control with the decisions I made. For many people, the years just after college are a stressful, exciting, and angst-filled time, and my experience was no exception. I had gone through some life changes: I'd left a startup company, ended a long-term relationship, and was actively trying to figure out what my version of being a "grown-up" would look like. Tracking my Life Satisfaction and Say to Do scores alongside my life decisions helped reveal whether my choices were actually moving me in a direction that led to greater happiness and success. You'll get a modified version of the Life Satisfaction surveys and Say to Do worksheets that I use myself at the end of this chapter.

As I made different choices over the next few years—got a new job, moved to a new city, started a new relationship—I was able to see how they affected my life satisfaction in specific areas and overall. Suddenly, I had a more objective way to make very subjective decisions. During a challenging patch in one relationship, I was able to look back at earlier times and see that this rough patch was probably an anomaly. When it came to work, I was able to compare my satisfaction at two previous jobs and tell that, in one, I had reported feeling more satisfied far more frequently. That was an early alert, and it helped me look more critically at what was starting to go wrong in this new job. By 2014, my tracking started to pay off. I had learned to set wiser goals that, when hit, actually corresponded to higher satisfaction levels. From 2014 to 2017, my Life Satisfaction and Say to Do scores began aligning with each other more—when one went up, the other did, too. Of course, when one went down, the other tended to as well.

Below, you can see that my Say to Do percentages contrasted with my satisfaction levels in my professional life from 2011 to 2017.

Professional Say to Do and Life Satisfaction Scores: 2011–2017

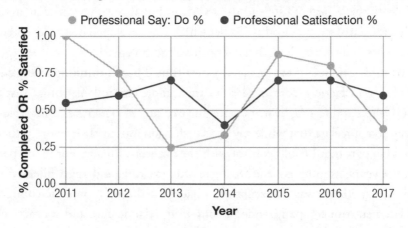

INTERPRETING YOUR FINDINGS

Of course, my interpretations of my personal data were not the only possible conclusions to draw. There are many factors that go into anyone's self-report of how they're feeling about a job or a relationship or even their level of satisfaction with the furniture in their living room at any given time. The point of the exercise was to remind me that I could and should make choices that attempt to optimize for life satisfaction. Otherwise, I realized, my life decisions might end up being thoughtless and random, perhaps driven by ill-informed efforts to please other people, perhaps driven by impersonal ideas of what some idealized person "should" do. Rather, tracking my Life Satisfaction and Say to Do helped me figure out what *I* should do. I could look at my graphs to see which successes had brought me joy, which failures had brought me grief, and make decisions built on personal knowledge.

Having an accountability partner helped me escape some of the echo chamber of producing data about my life and then trying to interpret it, too. My accountability partner has helped me notice patterns about my behavior that I didn't notice in myself. And one of the most gratifying things is to provide those same types of insights to my accountability buddy, too. For more on accountability buddies, check out chapter 5, "Organize to Motivate."

Tracking Your Life Scores

Below, you'll find a modified version of the Life Satisfaction survey and Say to Do worksheets that I use for myself.

Remember, this is for you; you are not competing with anyone else. These questions provide a way for *you* to gauge where *you* think you're at on your Life Satisfaction and Say to Do scores for the time frame specified. Just answer as honestly as you can. Grab a pen or pencil: it's time to reflect!

Self-Assessment of Life Satisfaction & Say to Do[8]

LIFE SATISFACTION SURVEY & SAY TO DO WORKSHEET		
Part I: Life Satisfaction Survey		
PICK YOUR TIME FRAME OF REFLECTION		
For this survey, please answer all questions with a specific time frame in mind.		
If you want to use your answers to this survey as a baseline or for assessing the effects of specific interventions during your daily self-experiments, choose "Last 24 hours."		
If you want to use it as a way to record your more general sense of your executive function over a longer time period, choose "Last 30 days" or "Last 3 months."		
LAST 24 HOURS ___ LAST 30 DAYS ___ LAST 3 MONTHS ___		

Personal
Ratings on a scale from 1 to 10:
   1 Never did it feel ideal/it felt miserable.
   5 Sometimes it felt ideal.
   10 Often/consistently it felt ideal.		

Work (School, Job, etc.)
Ratings on a scale from 1 to 10:
   1 Never did it feel ideal/it felt miserable.
   5 Sometimes it felt ideal.
   10 Often/consistently it felt ideal.		

	Personal	**Work (School, Job, etc.)**
1. Good Feelings		
a. You laughed a lot.		
b. You learned new things and felt stretched.		

c. You made time for breaks/experiences that refreshed your life.		
d. You left yourself open to adventure and potential growth experiences.		
2. Flow		
a. Your skills were a match for your challenges; your challenges felt manageable and interesting.		
b. You achieved flow — you felt so fully present, so focused on what you were doing, that you lost track of time.		
3. Relationships: Core and Network		
a. You were satisfied with your personal relationships (with your family and friends in your personal life, with your colleagues and professional network in your professional life).		
b. There was relationship growth; you felt positive when thinking about the future.		
For your personal life:		
a. You felt great about your primary relationships (a significant love interest, best friend, partner, etc.).		
b. This primary relationship was growing in a positive way; you felt hopeful about the future.		
c. If you were looking for a relationship, you felt like you made progress toward a positive outcome.		
For your professional life:		
a. You felt positively about whichever work relationship you feel is most important (your boss, teacher, mentor, business partner, biggest client, etc.).		
b. This relationship went in a direction you felt excited about.		
c. If you didn't have a primary work relationship, but you wanted to get into one (you were looking for a mentor or a new employer, you were applying to schools, etc.), you met or exceeded your goals for progress.		
4. Health & Appearance		
a. You felt healthy.		
b. You felt great about how your body looked.		
c. You felt great about how your body functioned.		
5. Physical Space		
a. You felt great about your home or work space physically.		

b. You felt great about your home or work space emotionally.		
6. Finances		
a. You were happy with the amount of money you made, or were hopeful about future earnings.		
b You were happy with the amount you saved.		
c. You were happy with how you spent your money.		
7. Wins		
a. You felt great about your day-to-day achievements.		
b. You hit or exceeded your goals.		
8. Time Management		
a. You felt satisfied with how you spent your time.		
b. Whatever time management system you used worked well.		
9. Connection to the "Big Picture"		
a. You felt connected — spiritually or otherwise — to something larger than yourself.		
b. You felt awe, gratitude, or compassion.		
10. Buokct List		
a. You made progress in your hopes, in your dreams, and in listening to your "inner voice."		
b. If you had a "bucket list" (things you want to do or experience before you die), you made progress on it during this period.		
11. Life Overall		
a. You felt satisfied with your life overall.		
b. You felt satisfied with your accomplishments.		
c. You felt satisfied with your progress toward goals.		
d. There was little you would change about how you were living.		

Part II: (Personal & Work) Say to Do Worksheets		
Annual Worksheet Answer these questions at the beginning of each year.	Personal	Work (School, Job, etc.)
Goal Record		
What percent of last year's goals did I hit? (Leave blank if this is your first year with Say to Do.)		

Reflection		
What goals or interventions worked? What didn't? What will you carry forward? (Leave blank if this is your first year with Say to Do.)		
New Goals		
Write down the goals you will hit by the end of this year.		
Hint: Use your answers to the Life Satisfaction survey as inspiration for what goals to choose.		
Example: I'll hand in a solid draft (I've edited it, my peers have edited it, it was fact-checked) of my book to my publisher by their deadline.		
Choose 3–5 goals total. Each goal should be SMART (Specific, Measurable, Achievable, Relevant, Time-based).		
Guideline: Aim to hit 70–80% of your goals.		
Score any lower than that and you may have aimed too high; score any higher than that and you may not have stretched yourself enough.		

Quarterly Worksheet Answer these questions at the beginning of each quarter.	Personal	Work (School, Job, etc.)
Goal Record		
What percent of last quarter's goals did I hit? (Leave blank if this is your first quarter with Say to Do.)		
Reflection		
What goals or interventions worked? What didn't? What will you carry forward? (Leave blank if this is your first quarter with Say to Do.)		
New Goals		
Write down the goals you will hit by the end of this quarter. Ideally, these are broken-down versions of your end-of-year goals. Hitting these quarterly goals will be milestones on the way to hitting your annual goals.		
Example: I'll finish the first draft of my book before the last day of the quarter.		
Choose 3–5 goals total. Each goal should be SMART (Specific, Measurable, Achievable, Relevant, Time-based).		
Guideline: Aim to hit 70–80% of your goals.		
Score any lower than that and you may have aimed too high; score any higher than that and you may not have stretched yourself enough.		

Weekly Worksheet Answer these questions at the beginning of each week.	Personal	Work (School, Job, etc.)
Goal Record		
The percent of last week's goals that I hit. (Leave blank if this is your first week with Say to Do.)		
Reflection		
What goals or interventions worked? What didn't? What will you carry forward? (Leave blank if this is your first week with Say to Do.)		
New Goals		
Write down the goals you will hit by the end of this week. Ideally, these are broken-down versions of your quarterly goals. Hitting these weekly goals will be milestones on the way to hitting your quarterly goals. Example: I'll draft a chapter by the end of this week. Choose 3–5 goals total. Each goal should be SMART (Specific, Measurable, Achievable, Relevant, Time-based). Guideline: Aim to hit 70–80% of your goals. Score any lower than that and you may have aimed too high; score any higher than that and you may not have stretched yourself enough.		

Daily Worksheet Answer these questions at the beginning of each day.	Personal	Work (School, Job, etc.)
Goal Record		
What percent of yesterday's goals did I hit? (Leave blank if this is your first day with Say to Do.)		
Reflection		
What goals or interventions worked? What didn't? What will you carry forward? (Leave blank if this is your first year with Say to Do.)		
New Goals		
Write down the goals you will hit by the end of today. Ideally, these are broken-down versions of your weekly goals. Hitting these daily goals will be milestones on the way to hitting your weekly goals. Example: I'll write 200 words by 4 p.m. today.		

Choose 3–5 goals total. Each goal should be SMART (Specific, Measurable, Achievable, Relevant, Time-based).		
Guideline: Aim to hit 70–80% of your goals.		
Score any lower than that and you may have aimed too high; score any higher than that and you may not have stretched yourself enough.		

TAKEAWAYS

1. Picking real-world outcomes will allow you to track whether your neurohacking has led to improvement in your broader life, not just an upgrade to a specific mental target.
2. Use your Life Satisfaction scores to assess whether your neurohacking experiments have improved your well-being.
3. Tracking your Life Satisfaction can offer you new insights. You can analyze a very simple measure of Life Satisfaction or a more complex one that tallies your satisfaction across many different life areas.
4. Improving your Say to Do score not only heightens your productivity, it enhances your relationships—strengthening how well you keep your word to yourself and to others.

PART III

Select a (Foundational) Intervention

Chapter 13

Placebo on Purpose

*"He who believes he can and he who believes he can't
are both usually right."*
—Confucius

Time Investment: 19 minutes
Goal: To understand the power of placebo and
harness it for your own purposes in your
neurohacking experiments—rather than
being fooled by it

Henry Beecher was a young American army doctor serving in North Africa and Italy during World War II.[1] Most of the military medical personnel were wholly focused on keeping as many patients alive as possible, and keeping themselves alive while they were at it. However, in efforts to ration the dwindling supplies, many medical personnel were administering only about half the recommended pain relievers and, in some cases, giving no pain relievers at all. Young Beecher noticed that one nurse, subject to the same shortages as the others, began injecting the patients for whom she could offer no actual pain relievers with an inert saline solution. When Beecher asked her why she did this, she explained that this fake medicine, when combined with her attentive and sympathetic care, seemed to help the ailing soldiers, who were none the wiser. As he watched, Beecher saw that the nurse was right; this unorthodox approach seemed to work. The patients reported that they felt better, and they even recovered at a surprisingly high rate.[2]

When Beecher returned to the United States, he analyzed the results of these types of substitute treatments — termed placebos — in dozens of different studies. On Christmas Eve in 1955, the *Journal of the American Medical Association* published Beecher's pioneering article "The Powerful Placebo."[3]

Today, Beecher's legacy lives on; the randomized, double-blind, placebo-controlled trial is the gold standard in drug testing. If the placebo is so powerful, any drug should prove its worth by besting the placebo in a double-blind test. The double-blind means that patients should not know whether they are receiving a real drug or a placebo, and the researchers also shouldn't know which patients are receiving the real treatments or the placebos.

In studies conducted over decades, placebos of various forms have shown impressive efficacy at relieving physical pain of all kinds.[4] More recently, researchers have been exploring ways in which other conditions, such as depression,[5] anxiety,[6] Parkinson's,[7] and severe arthritis[8] could be helped by placebo. In one study, researchers randomly told some sleep study participants that they had below-average or above-average sleep quality. The assigned quality of sleep, as opposed to the self-reported sleep quality, was a significant predictor of performance on several cognitive tests including verbal fluency and processing speed.[9] That is, participants who were told that they had a poor night of sleep, regardless of their actual quality of sleep, reflected patterns of cognitive functioning that would be expected if they had actually experienced a poor night of sleep. This study, among many, highlights why we're focusing on placebo in a book on cognitive enhancement.

The Science Behind the Placebo Effect

As a child, you may have heard the following rhyme: "Sticks and stones may break my bones, but words will never hurt me." As it turns out, multiple imaging studies of the brain show otherwise. The activity patterns that occur after emotional pain — especially feelings of social rejection or heartbreak — look remarkably similar to the patterns after physical pain.[10] The brain appears to be surprisingly indifferent to whether pain is caused by a physical injury or an emotion. Could the brain's willingness to equate experiences, emotional and physical, possibly even real and imagined, help you maximize your neurohacking efforts?

There seem to be a few different mechanisms that explain how placebos work; let's look at three of them. In the case of pain relief, the placebo triggers the release of endorphins, one of the body's own pain relievers,[11] through the brain's opioid system. Ever cut yourself badly and noticed that the pain dissipates after a while instead of continuing to burn? That's thanks to your body's homemade pain defenses. Placebos seem to trigger chemical messengers (neurotransmitters) in the brain, such as dopamine, that regulate reward, as well as those, such as serotonin, that play a role in regulating your mood, sleep, and appetite.[12] The dopamine release may help explain why Parkinson's disease has responded to placebos; Parkinson's disease is caused in part by the degeneration of neurons in the substantia nigra that release dopamine.[13] Therefore, the placebo-linked release of dopamine plays a role in countering the dopamine deficit that is at the origin of many of the clinical symptoms of Parkinson's disease, such as rigidity, resting tremor, and poor motor coordination. Depression seems to respond to placebos, too; this may be due to the role of serotonin. Some of the more effective antidepressant drugs work by boosting levels of serotonin in the brain (via blocking the uptake of serotonin in the brain as a selective serotonin reuptake inhibitor). Ted Kaptchuk and his Harvard Medical School colleagues who are on the forefront of placebo research have written about the possibility that placebos may help in depression by acting on that very same serotonin-mediated pathway.[14]

A more conceptual theory of why the placebo effect is so strong in humans is that placebos may be hacking our brain's natural prediction engines.[15] Ever hear someone else begin a sentence and find yourself completing it before they get a chance to? When asked how we knew what they were going to say, we may say something like, "I just know you" or "You've said something like that before." This sentence completion trick tends to happen more when we have heard the person speak frequently or we are familiar with their favorite topics and their natural speech cadences. Whether or not we are conscious of it, we are building generalizable predictive models for how that person talks.

Some scientists believe that this modeling and prediction engine that is constantly searching for patterns and generalizing from them is also what makes us especially susceptible to placebos. We look for patterns and anticipate the rewards at the end of an association sequence to the point where we don't even need to experience any external reward; merely letting our imagination prediction engines produces the same effect.

DOES THE FORM OF THE PLACEBO MATTER?

In 2016, I was fortunate to attend a talk by Ted Kaptchuk at Harvard. Kaptchuk emphasized that details matter for the placebo effect, and they can be culturally specific. In some countries—the United States, for instance—pills have a greater placebo effect than the use of biomedical devices. In some cultures, the greatest effect comes from injections. The white coat and the diplomas on the wall make a patient more likely to benefit from a visit. Clever pricing makes a difference. A study by Dan Ariely at Duke found that participants were more likely to experience placebo benefits from higher-priced pills than lower-priced pills.[16]

Reading more on Kaptchuk's work, I learned that even the color of the dummy pills used matters. Placebos that were red worked best for stimulants, white placebo pills were best for antacids, green was best for chilling out (anti-anxiety), and yellow was best for bringing on a sunny mood.[17]

PLACEBO CAN UPGRADE ATTENTION

Clothes can create a placebo effect too; in multiple studies, researchers found that clothing actually alters people's mental performance. Items studied most closely are uniforms, talismans (such as a lucky necklace pendant or bracelet), and a lucky sock. Oddly enough, these objects of superstition seem to really work—as long as the person believes in them. For example, researchers randomly assigned participants to complete a selective attention task. Some of them completed it while wearing a lab coat, and the others did not; those in a lab coat performed better. In a follow-up study, participants were randomly assigned to two groups; one was told that the white coat had belonged to a doctor, and the other was told that the coat belonged to a painter. The participants who were told the lab coat they were wearing was a doctor's coat performed better on the attention task than those who were wearing the exact same lab coat that was described as a painter's coat.[18]

Why might this be? Researchers found that participants believed that lab coats, the prototypical attire of scientists and doctors, were generally associated with being careful and attentive. In putting on a lab coat or the coat of a doctor, they may have found themselves thinking more the way they imagined a scientist or doctor would think: attentively and carefully. Perhaps if the researchers had asked the participants to perform a creative task, the group told they were wearing a painter's coat would have performed better.

THE PAIN YOU EXPECT

The placebo effect's evil twin, the "nocebo" effect, is ferociously powerful. Doctors have found countless examples of people believing themselves sick and then actually developing symptoms consistent with that sickness. In his *New York Times* bestselling book, *You Are the Placebo,* chiropractor Joe Dispenza describes a man who was diagnosed with end-stage cancer and died a few months later, just as predicted. This man, after having difficulties swallowing, went to a doctor who had diagnosed him with metastatic esophageal cancer, considered uncurable at that time. As recommended by the doctor, the man underwent an operation to have the cancerous tissue in his esophagus and stomach removed. The surprise occurred after his death. In the autopsy, his physicians discovered very little cancer. When additional doctors examined the initial scans that had led to the man's end-stage diagnosis, they were chagrined to find that the original doctor had made a mistake: what he thought was a cancer was actually an error on the image. It appeared that the man's body shut down because he, and everyone around him, believed that it would: "He died, quite simply, because everybody in his immediate environment thought he was dying."[19] In another study, Chinese people who had been born under a sign that was seen as inauspicious for good respiratory health had a far higher rate of death from a lung-related disease—but only if they endorsed beliefs about Chinese astrology.[20]

DOES EVERYONE RESPOND TO PLACEBOS?

As far back as "The Powerful Placebo," published in 1955, Henry Beecher recognized that some people responded to placebos and others did not. Since then, researchers have found that responders differ from non-responders in predictable ways; both psychologically and biologically. The psychological differences may not come as a surprise. Placebo responders were more likely characterized by personality traits such as openness to new experiences and a type of self-awareness that relates to awareness of your internal physical state (interoceptive awareness).[21] The biological differences were significant, too.[22]

Certain neuroanatomical and neurophysiological factors were predictive of an individual's response to placebo treatment. In addition, those with a genetic variant associated with higher levels of dopamine were more likely to benefit from placebo treatment in relieving their pain symptoms.[23]

You may be wondering whether you are more susceptible to being a placebo responder. A combination of factors plays into this. If you get your DNA sequenced using a consumer genomics product (such as 23andMe), you can actually look up whether you have one of the variants that may predispose you. For instance, if you look for the genetic marker rs4680 associated with the gene COMT (which affects dopamine levels), you could check whether you have the a/a variant (as opposed to a/g or g/g).[24] Of course, the placebo effect is complex and almost certainly not fully genetic. So, do not despair if you don't have the a/a variant on rs4680; it's still entirely possible that you'll respond well to placebos. The best way to find out is to try placebos for yourself. Which brings us to an important question.

CAN YOU PLACEBO YOURSELF?

For us self-experimenters, there's a big question here: Do placebos work only if you don't know you're getting them?

A number of studies have found that participants report improvements even when they are told that they are receiving a sugar pill or some other inert substance ("open label" placebo) — as long as they were also told about the body's ability to respond to a placebo. In 2010, Kaptchuk ran a study on irritable bowel syndrome (IBS) patients. Patients were either told that they were being given "placebo pills made of an inert substance, like sugar pills, that have been shown in clinical studies to produce significant improvement in IBS symptoms through mind-body self-healing processes" or they were simply given no treatment (but still had similar levels of attention by their medical provider). Astonishingly, the group that knew they were receiving a placebo reported a significant reduction in their symptom severity, better relief, and a trend toward better quality of life.[25] In 2017, a Harvard/Swiss group ran a larger study, this time on the effect of open-label placebo on heat pain tolerance; they compared people who were knowingly given placebos but with no rationale to those who received placebos with additional explanations such as the one that the previous group got about "mind-body self-healing processes" that result from a placebo. They found that individuals who received explanations fared better than those who received no explanation. Additionally, they found that those who were given a placebo with a rationale reported similar pain tolerance to those who received placebo thinking they were receiving an actual analgesic ("deceptive" placebo).[26]

The second hurdle is whether someone else must provide the placebo to you for it to work—such as a doctor in a white coat. I was not able to find rigorous studies comparing self-administered placebo to placebo administered by someone else (particularly an authority figure).

If you'd like to try a variation in which someone else give you the placebo, find someone who you respect and trust. It sounds silly but try having them administer your chosen intervention, and instruct them to say, "Clinical studies have shown significant improvement through mind-body self-upgrading processes." There are other types of placebo that certainly work well when self-administered; visualization is one of them.

Neurohacking with Placebos: Visualization for Upgraded Learning (and Performance)

One of my strongest exposures to the power of the mind-body connection to provide upgrades was in sports. Before my first match at the Junior Olympics (I competed in the racket sport of squash), I found a quiet spot away from where I would be competing. I closed my eyes. I mentally rehearsed exactly what I would do as soon as the door to the squash court opened. I imagined what my legs would feel like, what the crowd would sound like, how dry my mouth would taste, what the racket would feel like in my hands. Then, I saw myself serving perfectly, rallying with my opponent perfectly, recovering from setbacks perfectly. When the game began, I saw that my opponent was very strong. Her shots were accurate and powerful, and she was very fit. Very soon, she was winning. Between every rally, however, I returned to my visualization. I closed my eyes, watched what had just happened, identified what I was doing wrong, and then practiced playing it correctly. As the game wore on, I was able to find her weaknesses and take more and more of the shots I had visualized. As my mental game improved, my opponent's focus began to waver. She started cursing whenever she hit a bad shot; she rolled her eyes when the referee made a call in my favor. Now, all I had to do was wait for her mind to turn upon itself. It did. She began making self-inflicted errors. She ended up playing our final game at the bottom of her technical abilities—whereas I played at the top of mine. I won the match.

Many athletes use visualization. In fact, it's widely used among all types of performers. During his first years trying to make it in Hollywood, a young man wrote himself a $10 million check for "acting services rendered" to be cashed on Thanksgiving in three years. He tried out for acting roles all day every day; every night, he visualized. He parked on the famous Mulholland Drive, mentally practicing all the steps he believed would take him to his dream, really visualizing it. A few days before Thanksgiving three years later, he received a check for that amount for his role in *Dumb and Dumber*. The young man's name was Jim Carrey and he would go on to starring roles in *The Mask, Ace Ventura: Pet Detective,* and many others.[27]

Where else does visualization work? I've heard of cases of prisoners who spent their days of imprisonment imagining chess moves in their minds. Not allowed to read, write, or talk to anyone, Russian human rights activist Natan Sharansky used chess as a tool of survival, stating: "The KGB hoped that I would feel weaker and weaker mentally. Actually, I felt stronger and stronger."[28] Ten years later, his practice paid off: he beat world champion Garry Kasparov.[29]

In 1995, Alvaro Pascual-Leone of Harvard Medical School decided to investigate the effect of visualization on both behavioral performance and brain function. He compared the effectiveness of a group of practicing piano students who had access to a real piano to an ability-matched group who were instructed to practice playing but without the piano—simply imagining and moving their fingers over an air piano. A few weeks later, he compared the average playing abilities of the two groups. What he found shocked him. The visualization-only group showed significant improvement in their finger exercises (although not as good as the physical practice group). However, the brain changes in the visualization group looked similar to those in the physical practice group, suggesting that visualization alone appeared sufficient to change neural circuitry involved in this type of learning.[30]

What Jim Carrey, Natan Sharansky, and the piano-less pianists practicing with little more than their imagination all show us is the power of visualization and mental practice.[31]

Neurohacking with Placebos: The Growth Mindset as Intervention

Over the last few decades, Stanford professor Carol Dweck has become well known in both psychology and education circles for her research and efforts to share the benefits of possessing a "growth mindset." Growth mindset is not a placebo in the conventional sense of the word. However, it captures some of the most powerful aspects of the mind's ability to influence itself. In this sense, it shares a key ingredient of the placebo and, since our goal here is for you to learn how to use the placebo "on purpose," I consider it an unconventional vehicle for harnessing the power of placebo.

A person with a growth mindset believes that hard work is a prerequisite ingredient to success, or at least a critical ingredient in the process. Rather than being a signal that a person "doesn't have what it takes," people with a growth mindset believe that hard work actually provokes talent.[32] Because of this belief, people holding a growth mindset tend to be more persistent and resilient in the face of adversity. They tend to enjoy learning new things, because mistakes and failures are seen as a natural and necessary part of the learning process. Rather than fearing failure, they learn from it and actively seek new challenges.

In contrast, a person with a "fixed mindset" believes that potential is set; whatever talents you have are fixed at birth. A person with a fixed mindset believes there are smart people and not-so-smart people, and the not-so-smart people can't do much about their status, because intelligence is fixed. School is about discovering what your and other students' intellectual potential is; it is not about trying to improve upon it. Not surprisingly, when people with a fixed mindset discover that something is challenging, they will be more likely to give up and try something else. To them, any success must come fairly effortlessly, otherwise it is not a "real" success.

Growth mindset interventions have been aimed at changing attitudes in many areas, but a particularly striking one was directed at changing participants' ideas about personality, encouraging them to consider the idea that "personality is malleable." After a single, 30-minute session of computer-guided growth mindset training, a group of 12- to 15-year-olds reported significantly lower ratings of anxiety (over 12 percent improvement on average) and depression (over 26 percent improvement on average, as endorsed by their parents) at a follow-up nine months later than a matched group who had been randomly assigned to the control condition of a 30-minute session of computer-based supportive therapy.[33]

In Part V, you'll find self-experiments that leverage the power of the placebo that you can try at home. One is based on a study in which researchers told participants that cinnamon would make them more creative—and it did.[34] Another is based on a study in which researchers told participants that a nasal spray (or a pill) would help them control their emotions—and it did.[35] Before you jump into these self-experiments, though, let's survey a few more interventions. Next up is...sweat.

TAKEAWAYS

1. In medicine, a placebo is a procedure or treatment that looks like a real, physical treatment but instead acts psychologically. It can have both positive and negative effects; participants report the exact side effects they were told to expect, even when they received nothing but a placebo.

2. Placebos seem to work in two ways. The first may be by hijacking your brain's pattern-matching tendencies. The second is by hijacking your body's biochemistry.

3. Even when you know it's a placebo, it could still work. Multiple studies have shown that participants' symptoms improved even when they knew the treatment they received was "fake"—as long as they were told about the power of the placebo effect before.

4. The form the placebo comes in affects how well it works—including the color of the pill, uniforms worn, and the cost of the treatment.

Chapter 14

Sweat

"A sound mind in a sound body is a thing to be prayed for."
— JUVENAL (ancient Roman poet)

Time Investment: 17 minutes
Goal: To understand the power of exercise to upgrade your mental performance and learn how to incorporate it into a neurohacking experiment

Like many high school athletes, I faced a crossroads going into junior year: should I prioritize sports or schoolwork? To attract college coaches' attention, I would need to compete more outside of school to boost my national ranking. To have a shot at a competitive science program, I would need to succeed in the most advanced math and science classes at my high school. In both cases, I was starting from behind.

During the first two years of high school, both my athletic and academic performance had been erratic. One moment I was 100 percent focused, the next I was zoning out in la-la land. Not only did I forget homework assignments, I mentally checked out during important squash matches. Thankfully, two opportunities — one in science and the other in squash — presented themselves the summer before my junior year. A prestigious research laboratory invited me to intern, and an elite squash training camp in the UK accepted me as a trainee. I told Mark Lewis, my coach, that I'd heard that the level of play was much higher in the UK; I admitted I was scared. He nodded. "You probably should be. Here's how to prepare: Get in the best shape of your life."

He recommended a new regimen of high-intensity interval training on the bike and sprinting on the track. It was brutal. I combined this with my regular squash and weight training. By the end of the spring, I was astonished at how much stronger and faster I was. I also noticed something else: my head felt clearer.

At the squash training camp, I discovered that Mark was right. By the end of the first week, my muscles were screaming. But my training paid off; I didn't get injured. My summer research went similarly. For every day of the first week, my head was pounding by lunchtime. When my boss handed me a graduate-level microbiology textbook to study over the weekend, I put on a smile and said, "Sounds great!" Yet, as exhausted as I felt by the end of each day, I was always energized the next morning. Somehow, my physical fitness had spilled over into mental endurance.

That fall, I faced the hardest academic load I'd ever had. Additionally, the squash tournament season had me traveling almost every weekend. Yet while I was taking harder classes and had less time to study for them than before, my prior erratic academic performance—an A on one test, completely forgetting to hand in assignments before the next—was no longer a problem.

While this was far from a perfect experiment, I began to wonder: just how much could exercise alter brain functioning?

A Sweaty Science

Exercise may be one of the most magical cognitive interventions out there. From the research that's been done so far—on far more people than just me—it seems to work well on all four of our targets: executive function, learning and memory, emotional regulation, and creativity. Not only that, it starts working immediately. Even one exercise session can yield mental benefits—and in some cases, these benefits kick in before the session even ends!

How Does Exercise Affect Your Executive Function and Attention?

In 2016, neuroscientist Adele Diamond, a fellow of the Royal Society of Canada, and her colleague Daphne Ling published a review of interventions that specifically targeted executive functions.[1] They found broad gains from exercises that involved planning and concentration, like yoga, and from those that involved problem-solving or emphasized self-control, like Tae Kwon Do. Diamond and Ling also hypothesized that many competitive and team sports would boost executive functions, too, but they couldn't find high-quality research studies directly assessing them. A 2014 randomized controlled trial investigated whether an after-school exercise program called FITKids would help over 200 preadolescents with their executive function. The researchers found significant improvements in inhibition and flexibility, as well as promising changes in their brains' electrical activity.[2] I would imagine that high-level competition at a sport like squash would yield similar cognitive benefits, but I'm obviously biased; we need more research.

Can You Get Smarter from Just One Exercise Session?

In 2012, Yu-Kai Chang and his colleagues at the University of North Carolina–Greensboro conducted a meta-analysis of 79 studies to understand the acute effects on cognition of one session of exercise.[3] These studies included more than 2,000 research participants drawn from all over the world, including women and men, children and older people, healthy and impaired individuals. Studies included all intensity levels of exercise, from very light to maximal effort and different types of exercise: aerobic (cardio), anaerobic (sprinting), resistance (strength training), and combinations of aerobic and resistance training. His team looked at different types of cognitive abilities, including measures of both executive function and memory.

Chang and his colleagues found that, across the board, a single bout of exercise had small but positive effects on cognition. These findings held true for measures taken during exercise, following exercise, and a delay after termination of the exercise session. Attention, specifically concentration, was one of the types of cognitive outcomes most impacted by exercise: exercise increased concentration-based task performance more than almost all other cognitive abilities that the team measured.

When it came to executive function abilities and memory—how participants performed on tests involving decision-making, how fluidly they could express themselves verbally, how well they remembered stimuli, and how well they could inhibit incorrect responses and select correct responses—participants gained between 15 and 20 percent.[4] As expected, different brains respond differently, so let's dive into some sources of variation across people. In particular, we'll explore eight areas that Chang and his colleagues (as well as other research groups) further probed. As you read the following questions, try to predict the researchers' findings.

WHICH IS BETTER AT PRODUCING INSTANT ENERGY: EXERCISE OR CAFFEINE?

In 2017, an American research team was curious what would help a group of sleep-deprived college women (who, in this study, averaged less than 6.5 hours of sleep per night) stay awake. They had each woman exercise for 10 minutes (by walking up and down stairs) or take an amount of caffeine equivalent to half a cup of coffee or take placebo pills. The stair walking improved their energy the most in the short-term.[5]

WHO TENDS TO BE MORE CLEAR-HEADED — THE COUCH POTATO OR THE TRAINED TRIATHLETE — AFTER THE WORKOUT ENDS?

Based on categorizations set by the American College of Sports Medicine for low, moderate, or high fitness levels, average cognitive performance after exercise improved by more than 12 percent for fit participants and more than 8 percent for people of a moderate fitness level.[6] The low fitness group's experience was mixed; some actually performed worse cognitively than they had before they exercised, but others improved by nearly 23 percent over their pre-workout levels. The lack of benefit in one part of the lower fitness level could possibly be attributed to individuals being too tired out by the exercise to benefit immediately. Overall, exercise appears to benefit people of varying fitness levels, and these benefits seem to be most immediate in fitter individuals. This trend that fit individuals may exhibit more observable cognitive benefits is consistent with findings in Chang's meta-analysis.

HOW DOES THE INTENSITY OF THE EXERCISE AFFECT COGNITION AFTER EXERCISE — DOES A HIGHER OR LOWER INTENSITY WORKOUT HELP YOU THINK BETTER?

People's performance on cognitive tests after a workout tended to be better than it was before for almost any intensity of exercise. According to the

American College of Sports Medicine guidelines, levels of exercise intensity relate to the percentage of a participant's maximum heart rate that has been achieved. After light and moderate exercise (50 to 76 percent of max heart rate), average cognitive performance improved by 8 percent. For hard exercise (77 to 93 percent of max heart rate), it was around 12 percent better on average. For very hard exercise (greater than 93 percent of max heart rate), the reward was even bigger—around a 16 percent average improvement! The only exercise intensity that didn't help was very light exercise (less than 50 percent of max heart rate); here, average cognitive performance went down by about 4 percent.

HOW MUCH DO YOU NEED TO SWEAT?

One of my favorite sections in the book *Spark,* written by psychiatrist and Harvard Medical School professor John Ratey and science writer Eric Hagerman, is about the biology at play when you walk, jog, or sprint.[7] It delves into the effects of intensity across similar movements:

Walking: a stroll toward satisfaction

Ratey and Hagerman describe how walking at 55 to 65 percent of your maximum heart rate is a way to enjoy better mood. At this low intensity of workout, you are in a fat-burning mode, which increases the amount of free tryptophan in the bloodstream. Not only is tryptophan a precursor to serotonin, the neurotransmitter widely credited with providing feelings of satisfaction and contentment, but tryptophan also alters how norepinephrine and dopamine get distributed. Combine the increase in serotonin with norepinephrine's role in attention and dopamine's role in motivation, and you can see why our ancestors' brains evolved to enjoy walking.

Jogging: stress now, worry less later

Jogging—or exercising at a moderate intensity, 65 to 75 percent of your maximum heart rate—brings unique benefits of its own. Moderate exercise is basically stress training. Physical resilience is enhanced, because adrenaline and cortisol got pumped into the bloodstream during the moderate exercise. Your body jumps to regulate those as part of its stress response, and the systemic regulator of that is the hypothalamic-pituitary axis. Brain-derived neurotropic factor—or, as Ratey likes to call it, the brain's "Miracle-Gro"—is a building block for new neurons, and it jumps in to strengthen neural circuits, too. Atrial natriuretic peptide is a peptide hormone that is produced by the heart muscle and also helps moderate stress. Endorphins and endocannabinoids decrease pain and increase feelings of calm.

Running or sprinting: pain now, big brain later

In a high-intensity workout—say, when you're going at 75 to 90 percent of your maximum heart rate—your body assumes you are in an emergency mode. When you go into an anaerobic zone—using energy stored in your muscle instead of just what's available in your bloodstream—your brain's pituitary gland triggers the release of human growth hormone (HGH). HGH tends to decrease with age, but it plays a key role in increasing brain volume, managing many of the growth factors mentioned above, balancing neurotransmitter levels, and encouraging neuron growth. HGH can stay present and working hard in the bloodstream for hours after you finish your workout. In *Spark,* Ratey and Hagerman describe a research study at the University of Bath, England, in which a single 30-minute sprint on a bicycle led to a 600 percent increase in HGH that kept building for two hours after the sprint.

TO MAXIMIZE THE COGNITIVE BENEFITS OF A SINGLE SESSION OF EXERCISE, SHOULD YOU DO CARDIO, RESISTANCE TRAINING, OR A COMBINATION?

Chang and his colleagues found that combining aerobic and resistance training led to the largest effects, a roughly 12 percent immediate improvement in average cognitive performance. In contrast, aerobic exercise alone led to a gain of only about 4 percent on average, and resistance alone actually led to an average drop of about 12 percent.

So, what are the takeaways from these findings? When you're in need of a cognitive boost from exercise, measures of cognitive benefits across several studies show the greatest increase comes from a combination of strength and aerobic exercise—like the 15-minute protocols you'll read at the end of this chapter. However, these findings are averaged across people and across studies.

Like every intervention we'll discuss, your brain's response may be different from that of the people mentioned in the studies—so try a few approaches and test to see what works best for you! Also, even if you find that one type of exercise does not yield the cognitive benefits you were hoping for, remember that different forms of exercise have lots of other health benefits extending far beyond cognitive performance. For example, health benefits of resistance training include helping to prevent osteoporosis and enhancing cardiovascular health.[8] In a 2010 review conducted out of the University of Georgia by Kate Lambourne and Phillip Tomporowski, cycling won out against running. In their review of more than 40 research studies on exercise and cognition, they found that "cycling was associated with enhanced performance during and after exercise, whereas treadmill running led to impaired performance during exercise and a small improvement in performance following exercise." Specifically, the average cognitive boost measured following running was around 4 percent, with some getting as much as 8 percent. For cycling, it was around 8 percent, with some participants gaining as much as a 12 percent boost.

AGE: DOES A SINGLE SESSION OF EXERCISE AFFECT THE COGNITIVE PERFORMANCE OF PEOPLE OF DIFFERENT AGES DIFFERENTLY—AND IF SO, HOW?

The most significant positive effects for cognition from a single bout of exercise went to three different groups: high school–aged kids, adults over 30 years old, and older adults (over 65 years old)—for these ages, exercise was estimated to be providing a 4 to 8 percent boost to average cognition. Elementary school kids and young adults (20 to 30 years old) still enjoyed cognitive benefits from exercise, but they were estimated to be in the range of 2.5 percent on average. Keep in mind that these estimates are averaging across many, many studies, though, so the effects for individuals were much smaller in some cases and far larger in others.

TIME OF DAY: WHAT TIME OF DAY SHOULD YOU EXERCISE TO IMPROVE YOUR COGNITION THE MOST?

Chang and his colleagues found that exercising in the morning was associated with an average cognitive performance boost of nearly 16 percent. The afternoon and evening times were not statistically significant, perhaps because there was too much variability in the results. Afternoon exercise studies averaged a cognitive boost of about 4 percent, and evening studies averaged out to a drop of around 8 percent. While I encourage you to try morning workouts yourself, I'll admit that I did not. Given that my typical bedtime while writing this book was around 3:15 a.m., a morning workout sounded like torture to me.

How Does Exercise Affect Your Learning and Memory Ability?

In high school, while trying to squeeze in extra workouts, I started doing calisthenics during my study breaks. After the second heart-pumping break, I noticed that studying came more easily. When I told a friend, she decided to run an informal experiment with me: How well did we study after a break where we did jumping jacks and push-ups versus a break where we just stood up from our chairs and stretched? Word spread as the answer revealed itself. Pretty soon, there was a whole group of us blasting music in the girls' bathroom, dancing in front of the mirrors, doing jumping jacks, and then, after our timers went off, racing back to our desks to reap the benefit of our newly sharpened minds.

Chang and his peers found that when it came to performance on tests of cognitive skills that were already well learned, such as addition and subtraction, exercise boosted average performance by about 8 percent. When it came to new information, exercise improved people's free recall (such as a list of 15 new vocabulary words that you are given time to study) by nearly 19 percent. Some participants' free recall scores improved by around 23 percent.

At a study at the University of Muenster, Germany, a group of healthy young men were randomly assigned either to relax and sit down for 15 minutes, to sprint for three minutes at increasing speed two different times (with slower running in between), or to run at a low, steady pace for 40 minutes.[9] The sprinter group learned vocabulary words 20 percent faster right after their run. They also had higher increases in brain-derived neurotrophic factor (BDNF, aka brain "Miracle-Gro"). Sprinting also seemed to elicit higher dopamine levels, which was associated with better intermediate retention (remembering the vocabulary words a week later). Sprinting also elicited higher levels of epinephrine, associated with longer-term retention of the vocabulary words (after 8 to 10 months). The fact that the learning difference persisted even many months out surprised me. What if intensely exercising right after learning something new had no effect but intensely exercising a few hours afterward consolidated memory more effectively? In one study, 72 participants learned the association of pictures with specific locations and then were randomly assigned to do 35 minutes of exercise right away (high-intensity, up to 80 percent of their max heart rate), or to wait for four hours and then do the 35-minute high-intensity exercise, or to do a control protocol where they simply watched nature videos. Forty-eight hours later, all three groups were quizzed to see how well they remembered the picture-location associations. The group that exercised 4 hours after studying performed the best.[10]

How Does Exercise Affect Your Emotional Self-Regulation?

Self-regulation is about controlling your emotions and behavior. Over the long term, exercise seems to improve self-regulation. Researchers at Macquarie University, Australia, had participants track their behaviors over a four-month period.[11] After two months, participants started a regular exercise regimen. As compared to their two-month "control phase," the participants enjoyed significant improvements in numerous aspects of their ability to self-regulate. They reported decreases in stress, emotional distress, smoking, and alcohol and caffeine consumption, and they ate more healthily, controlled their emotions more, did their household chores more, attended to their commitments more, improved their study habits, and monitored their spending.

How Does Exercise Affect Creativity?

Exercise also seems to increase energy,[12] decrease negative feelings, and boost positive feelings. Researchers from American and Canadian universities found that 10 minutes of exercise seemed to decrease negative emotions.[13]

Since we know that mood is boosted by exercise, and since positive mood is one of creativity's better predictors,[14] it's not much of a surprise that creativity also goes up with exercise. We know exercise improves mood and mood improves creativity...so is exercising just helping creativity via mood-boosting?

A group of researchers from Middlesex University in the UK compared the performance of participants on divergent thinking tasks after they had watched a "neutral" video versus after aerobic exercise. They found that exercise enhanced a measure of creativity specifically related to one's ability to produce a variety of responses, *independently* of mood.[15] As for what types of exercise were likely to boost creativity, researchers at Stanford found that walking had an acute effect on both convergent and divergent thinking. They tested this by comparing people sitting, walking on treadmills, and walking outdoors. They found that walking outside produced the most novel and highest-quality analogies.[16]

A team of researchers at the University of Graz, Austria, found that people who exercise regularly may generate creativity more spontaneously, whereas the creative ideas of sedentary folks arise predominantly from exerting concerted efforts.[17] This study showed that both frequent exercisers and sedentary folks could achieve high creativity. It also showed that high creativity could come about both when we least expect it—like the classic "eureka!"—and when we are actively trying to be creative.

TAKEAWAYS

1. Exercise can boost cognitive performance. Immediately after light and moderate exercise, average cognitive performance improved by about 8 percent on average. For hard and very hard exercise, the average cognitive boost was around 12 to 17 percent.
2. Right after resistance training, cognition dropped, but when resistance training was combined with aerobic exercise, cognition was boosted. Cycling bested running in its immediate cognitive boosting effects.
3. Exercise affects the cognitive performance of people of different ages differently.

4. The time of day of exercise can have an impact. Morning exercise boosted average cognitive performances by nearly 17 percent. Afternoon exercise boosted them by about 4 percent. Evening exercise dropped them by around 8 percent.
5. Exercise can have an effect on each of our four target areas: executive function, learning and memory, emotional regulation, and creativity. It can also lift mood and fight sleep deprivation.

Chapter 15

Let There Be (Blue) Light

> **Time Investment:** 12 minutes
> **Goal:** To understand the power of blue light to
> upgrade your mental performance and learn how to
> incorporate it into a neurohacking experiment

A SAD Story

Despite having lived in snowy Boston for most of my life, it apparently took living in sunny California for me to get SAD—seasonal affective disorder, that is. In January 2019, a new long-term work project enabled me to do all my work during the afternoons and evenings. I could stay up as late as I wanted and then sleep from dawn to early afternoon. As a proud night owl, I considered this a heaven-sent gift.

Although I didn't immediately connect the events to each other, a few weeks after starting my full "vampire schedule" (as my mother liked to call it), I started to feel... different. My friends' jokes stopped making me laugh. My to-do lists began behaving oddly; by end of the day, there were more unfinished items remaining than there had been at the day's start. I kept catching myself staring at the wall, only to check the clock and realize I'd been doing that for many minutes. Then, there was the fact that I was sleeping way more than normal. When I suddenly found myself crying in the middle of the afternoon with no idea why I felt so sad, I had to admit something was wrong.

A depression test called the CES-D (Center for Epidemiologic Studies Depression scale) scored my symptoms at just below the "severe" level. On the phone with my dad back in Boston, I shared my symptoms, and he asked an unexpected question: How much sun was I getting?

My dad suggested a solution: a light therapy lamp. After our conversation and a lot of research, I ordered a $40 lamp that gives off 5,000 lux (one lux is the illumination of a single candle about three feet away; direct sunlight is around 100,000 lux). Even at 5,000 lux, I directed the lamp toward the side of my face, not right into my eyes,[1] and I placed it two feet away. The effects were fast. I started thawing from the inside out within the first couple of sessions. For the next few weeks, I would spend the first half hour (or sometimes full hour) of my workday with the little square of light on the side of my face. Soon, jokes started to seem funny. The random urges to cry slowly stopped. My to-do lists started getting done. Just a week after starting the light therapy, I took the CES-D again: I was safely back in the "nondepressed" range. Hallelujah!

While light therapy is not going to be the perfect depression treatment for everyone, my curiosity was certainly piqued. How had light wielded this magic upon my brain? And could it do more than just improve SAD? After all, it had buoyed my concentration and productivity, too. Could light enhance cognition more generally?

The Science of Light Therapy

Researchers have been able to detect how alertness and mood get regulated by light using MRI scans. As soon as light is detected by the brain, a myriad of brain regions come to life, including regions that regulate alertness (such as the hypothalamus and thalamus), as well as memory and emotions (such as the amygdala and hippocampus).[2] Some of the key players in our light drama are photoreceptors (neurons that are specifically activated by light) in the eye and two chemicals: an alertness-causing chemical called norepinephrine and a sleepiness-causing hormone called melatonin.

Here's how they work together: When light enters your eye, photoreceptors send electrical signals back into the brain. A nearby region of the brain (the suprachiasmatic nucleus) receives the signals and subsequently decreases production of melatonin. The story is roughly: more light, less melatonin; less melatonin, less sleepiness. Conversely, when there's less light, there's going to be more melatonin produced, and with more melatonin comes more sleepiness.

During the day, however, norepinephrine plays a bigger role. Photoreceptors in the eye also trigger a different nearby brain region (the locus coeruleus) to increase the release of norepinephrine throughout the cerebral cortex. If you're familiar with the fight-or-flight response, norepinephrine plays a big role in getting you wide awake and ready to act. So, more light also means more norepinephrine, which means more alertness.[3] Basically, sleepiness (regulated by melatonin) and vigilance (regulated by norepinephrine) are on a balance beam that is regulated by the presence of light.

Light, Suprachiasmatic Nuclei (SCN), and the Pineal/Melatonin Circuit[4]

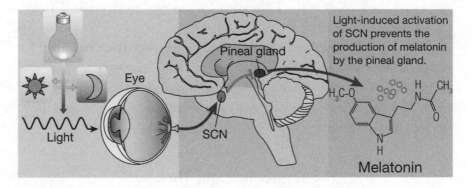

THE COLOR OF LIGHT, AND A NOTE ABOUT SAFETY

The color of light affects how the brain responds to it. We're warned against staring at computer screens late at night for the exact same reason that going out in the sun is one of the best cures for jet lag: blue light. The sun is our natural source of blue light, and we have evolved to associate the amount of blue light with the time of day. Consider the warm, orange glow of sunset in contrast to the pure blinding light of midday. Our circadian rhythm is tuned to these color-based changes. Staring at a computer screen—which emits primarily blue light—late at night is a major cause for insomnia; the brain "thinks" that it is midday and so doesn't fall asleep. For this reason, for the final hour before I go to bed, I wear glasses with orange lenses that block blue light (that I bought for $10 online). In addition, I use a software program that gently shifts the balance of my computer screen light from short wavelength (blue) to long wavelength (yellow and red) as the day wears on. By bedtime, my computer screen is tinted amber. There are plenty of software programs that do this; I've used a free program on my laptop and phone for years. Sometimes you can even set this color shift timer in settings.

Some people are especially sensitive to blue light and should be careful with the interventions described in this chapter. For instance, there have been reports that light therapy can trigger mania or hypomania in people with bipolar disorder (summer months—times of increased light—tend to be times with greater risk of manic and hypomanic episodes). Even among people not specifically diagnosed with bipolar disorder, some workers have reported becoming more irritable when subjected to blue light for hours at a time.[5]

Another group of people that should be extra careful with blue light are those with diagnosable eye conditions or diseases or people at risk of eye diseases, such as diabetics, who may experience eye strain when exposed to too much blue light.[6] People often forget to blink when they are in front of a computer screen, so reminding yourself to blink frequently is a good idea regardless. A good rule of thumb to avoid eye strain: use the blue light when another light is on, too.

Now let's dig into the cognitive enhancement part. Does it only work for people struggling with SAD? Also, does light work as a general cognitive enhancer—like caffeine, for instance? If so, does it work better, worse? A handful of randomized controlled studies address my questions.

How does blue light affect cognition generally?

A group of researchers from the University of Bordeaux, France, sent participants on an epic cross-country road trip late at night.[7] They wanted to know: Would blue lights on a car's dashboard improve driving accuracy? The 48 brave participants were healthy (that is, not depressed) men ranging from 20 to 50 years old, and they drove 250 miles in the middle of the night on three separate occasions. The men were randomized to one of three groups: one group was given caffeine (two doses of 200 mg of caffeine, the equivalent of about four cups of coffee), one group was exposed to continuous blue light, and one group was given a placebo (fake caffeine pill). The findings? The placebo group struggled to stay awake, each man on average crossing over the road lines about 26 times during his nocturnal adventures. In contrast, the caffeine and blue light groups performed nearly twice as well as the placebo group; on average, each man in the caffeine group crossed out of his lane about 12 times. The blue light group did almost as well as the caffeine group, with only 14 crossovers on average.

WHICH COLOR LIGHT IS BEST?

In the previous study, the researchers chose blue light (a short wavelength around 480 nanometers [nm]), but our ancestors evolved to respond to the sun's light, which combines all colors into a bright white (ranging from 400 to 700 nm). As I read through the studies, one other thing puzzled me: Why was the protocol for my SAD light therapy optimized around exposure to 10,000 lux, but the blue light studies all involved around 200 lux of exposure? A popular manufacturer of blue lights, Philips, explains that blue light "is the most efficient kind of energizing light, requiring just 200 lux to achieve an effect similar to 10,000 lux of white light."[8] This may be due to the fact that researchers have found photoreceptors in the eye are more sensitive to blue light than to white light. So, focusing solely on the most relevant part of the spectrum packs just as much punch as the whole spectrum.

In general, blue seems to be the most effective color for enhancing mental performance. In an observational study of more than 100 British office workers, a comparison was made in self-reported surveys of the workers under white lights versus blue-white lights. The workers reported significantly increased daytime alertness and better work performance under the blue light than the blue-white light.[9]

HOW DOES BLUE LIGHT AFFECT EXECUTIVE FUNCTION AND ATTENTION?

In 2013, a Swedish randomized controlled experiment found that blue light improved performance on executive function tasks. The researchers recruited a group of 21 healthy participants and subjected them in turn to blue light, white light, caffeine, and placebo caffeine pills as they performed a range of cognitive tasks on different days. The result? The participants performed better cognitively with blue light and caffeine than with white light or place-bos. While both caffeine and blue light improved their performances on a visual reaction test, they performed better on a test of executive function after blue light than caffeine.[10] Oddly enough, blue-eyed participants were especially aided by the blue light exposure. This may be due to the fact that blue eyes have fewer layers of protective pigment than do brown eyes, so blue-eyed people tend to be more sensitive to light than their darker-eyed counterparts.[11]

How long do blue light's cognitive enhancement effects last? Most of the studies tested participants during the same time that they were being exposed to light or shortly after light exposure. What happens when you do light therapy, wait, and test after a delay? In 2016, researchers at the University of Arizona and Harvard Medical School investigated that question for working memory, an aspect of executive function. They waited 40 minutes after their participants had a 30-minute exposure to either blue or amber light (for the control group). Then, the researchers looked at how participants performed on a working memory test and at what brain regions were active during the test.[12] In the functional MRI (fMRI) scanner, the participants in the blue light exposure group showed increased activation in areas related to working memory (such as the left and right dorsolateral prefrontal cortex). The blue light exposure group also performed faster on the hardest of the working memory tests. While more research is needed—I didn't find rigorous studies investigating more than 40 minutes after light exposure—these studies imply that the cognitive effects of blue light do not fade immediately.

How does blue light affect memory?

Does blue light help with memory tasks? A study conducted in 2017 on 30 healthy 18- to 32-year-olds compared the effects of blue light to the effects of yellow or amber light (around 580 nm).[13] The University of Arizona research-ers compared the memory performance of individuals randomized to 30 minutes of blue or amber light. Participants were asked to learn a list of words and immediately recall as many as they could remember. Then, after a delay of 90 minutes that included a light washout (to undo the effects of indoor lighting) followed by exposure to either blue or amber light, participants had to recall the list. The result? On average, the blue light exposure group forgot less than 2 percent of the words they had learned, whereas the amber light group forgot almost 15 percent of the words they had learned. Blue light wins again.

TAKEAWAYS

1. Light affects alertness in the brain; blue light had similar effectiveness to caffeine when study participants attempted to drive while sleepy. The effects of blue light compare favorably to the effects of caffeine.
2. Blue light enhances the brain better than other colors. Blue light worked better than white light to reduce daytime sleepiness and improve work performance. Blue light worked better than amber light to improve memory performance.
3. Blue light has minimal side effects for people *without bipolar disorder or eye conditions*. Like caffeine, however, it can lead to excessive alertness. Avoid using it right before bed or if you have been diagnosed with a mood disorder. Anyone with eye conditions should ask their doctor before using it.
4. How long do the effects last? There is still limited research, but at least one study showed that the cognitive enhancement effects persisted 40 minutes after the light exposure.

Chapter 16

Rewriting the Brain's Signature

"To me, thought-controlled computing is as simple and powerful as a paintbrush....I look forward to the day that I can sit beside you, easel-side, watching the world that we can create with our new toolboxes and the discoveries that we can make about ourselves."

—ARIEL GARTEN

> **Time Investment:** 20 minutes
> **Goal:** To understand the power of neurofeedback to upgrade your mental performance and learn how to incorporate it into a neurohacking experiment

As a child, Abhinav Bindra writes, he did not look like any kind of Olympic hopeful. In his autobiography, he described himself as a "fat boy [who] hated physical activity and was ambivalent about playing sport." Once a month, however, he watched his father clean his gun. This routine captured his interest. At the age of 10, his father allowed him to shoot the gun. In that moment, Bindra's passion for shooting was kindled.[1] He was the youngest athlete at the 1998 Commonwealth Games, but he came in only seventh at the 2004 Athens Olympics. By then, he was almost 22 and probably realized that his chances at a gold medal were disappearing. Over the next four years, he tried many self-improvement experiments—including rock climbing and drinking yak milk. Most interesting to me, however, were Bindra's experiments with biofeedback.

With his sports psychologist, Bindra focused on learning to control his breathing and heart rate, controlling whether or not he was in fight-or-flight mode (shooting can be successful in either mode, but switching from one to the other causes errors), eliminating excessive muscle tension, reducing his "internal monologue," and generally improving his reaction time and focus.[2] The upshot? Bindra won the gold at the Beijing Olympics in 2008 for the 10-meter air rifle. It had taken 112 years, but India had finally earned its first gold medal in an individual sport.

What Is Biofeedback?

Almost 10 years after Abhinav Bindra won his gold medal, I drove to Haight-Ashbury in San Francisco to meet with a biofeedback pioneer, George Fuller von Bozzay, a former University of California–San Francisco professor. He agreed to discuss my research into the topics covered in this book, but when I shook hands with the doctor, I immediately apologized when I saw him wince. "My hands are always cold," I explained hastily.

"Perhaps you suffer from anxiety?" he suggested. I felt my face heating up. Nothing like being told that you are nervous when you are trying very hard to come across as professional. I assured him again that my hands are always cold. He nodded. "Perhaps you are always anxious, then." Wow, this was going well.

"Let us try an exercise." Fighting back skepticism and a little embarrassment, I listened as he outlined his idea. He wanted me to do a progressive relaxation meditation. "You are here to learn about biofeedback, correct?" I nodded. He handed me a fingertip temperature monitor. "The best way to learn about it is to experience it. Now, measure your fingertip temperature." I did so. "During the meditation, watch to see if it changes." My face probably conveyed some dismay. "Just watch—who knows what will happen? When we finish the meditation, you can check your fingertip temperature again."

After 10 minutes of the meditation, I had to admit that I felt much more relaxed. Then, I looked down at the fingertip temperature monitor. I did a double-take. My finger temperature had risen by over 10 degrees Fahrenheit. I looked around the room—maybe he had increased the room temperature during the meditation? Maybe it had been cold outside and I'd just been sitting in his office for so long that my fingers had naturally heated up. Had I just altered my own body through the power of thought alone?

After thanking Dr. Fuller von Bozzay for his time, I drove home and immediately ordered my own device (for less than $15 online). There had to be more to it than that. I'd heard of Buddhist monks—through sheer concentration and deep meditative practice—drying wet sheets placed on them in a cold room, but I assumed you needed years to perfect that level of ability.[3]

When the fingertip thermometer arrived a week later, I replayed the same 10-minute meditation. It happened again: my fingertips started out deathly cold but, by the end, they had risen to the level of most normal humans. No change in external room temperature, either, I checked for that. It was just me, the meditation, the temperature monitor, and some magical change to my body via my thoughts. Disbelieving, I tried it again. And again. I tried different times of day and in different circumstances.

It wasn't perfect. Sometimes, I couldn't concentrate on the meditation at all. Then, the temperature didn't change as much. Other times, I was already pretty relaxed when I started, so my fingertip temperature didn't have as far to rise. Still, I succeeded frequently enough that my skepticism subsided. This mind-body connection thing was apparently for real.

What about doing just the meditation without watching the fingertip monitor throughout? My initial temperature rose, but it wasn't as effective as when I intermittently checked the monitor. Using the fingertip temperature monitor seemed to ground me in what my current internal state was. That allowed me to use the meditation more effectively.

"And that is how biofeedback works," von Bozzay said the next time I saw him. He explained that there are many different modalities of biofeedback; in addition to temperature, you can watch your heart rate change and learn to change that. You can measure the number of times you breathe in a minute and learn to tie that to an emotional state. We tend to breathe shallower and faster as we get excited, and more deeply and slowly as we relax.

Von Bozzay explained that I could also watch or listen to my brain waves and learn to control them, just as I had learned to control my fingertip temperature. "It will be harder and take longer, though," he cautioned. "When you use a monitor to detect brain waves, that type of biofeedback is called EEG biofeedback—or, some call it neurofeedback."

The Science Behind Neurofeedback

While using EEG is the oldest and most common form of neurofeedback, newer versions rely on detecting changes in blood flow and thus require fMRI or other metabolic imaging equipment, such as the far less expensive and more portable hemoencephalography (HEG) device. But, since EEG-based neurofeedback currently remains the most popular, I'll focus on that.

While the human brain shows all different brain waves at any given time, when there is more of a particular brain wave, it is often associated with a particular mental state. Alpha waves, a brain wave that runs 8 to 12 cycles per second, are associated with feelings of relaxation and calm focus, and increasing them is often recommended for people who feel stressed and anxious.

EEG Brain Wave Activity[4]

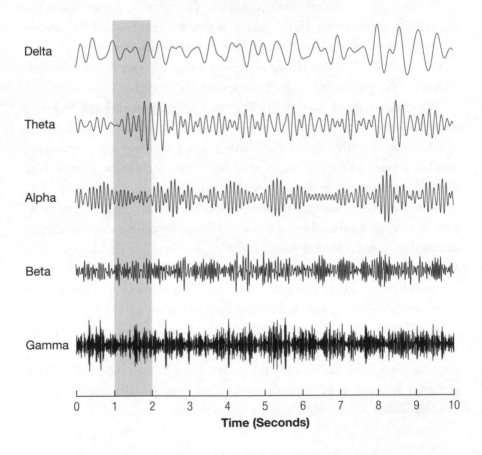

Producing disproportionately more beta brain waves is associated with being in a state of alertness and problem-solving; children exhibit more beta waves when working on math homework than they do while watching a TV show, for instance. More gamma waves are associated with deep learning and creativity.

THE SAFETY OF NEUROFEEDBACK AND PROTOCOL CONSIDERATIONS

Side effects from neurofeedback are not common. However, neurofeedback users occasionally report fatigue, headaches, nausea, tics, and sometimes nightmares. It's also worthwhile to be aware of reversal protocols. This occurs when a researcher or clinician ends up creating a feedback protocol that increases (rather than decreases) atypicalities in the participant's brain waves. For instance, ADHD is often characterized by an excess of theta brain waves; thus, effective neurofeedback protocols often focus on decreasing theta waves. A reversal protocol, however, would do exactly the opposite—it would *increase* theta waves. As a result, ADHD symptoms would likely increase. The authors of a 2007 article on reversal protocols cautioned neurofeedback practitioners to personalize treatment rather than overgeneralize, especially if a client has more than one disorder going on at the same time. For instance, while many alcoholics have an EEG profile that includes excess beta waves, not every alcoholic has this profile. In fact, the paper stated that one in four alcoholics have ADHD in addition. For this group, decreasing the amount of beta waves would exacerbate their ADHD symptoms further, so care must be taken to calibrate neurofeedback correctly.

THE HISTORY OF NEUROFEEDBACK

In the late 1960s, Joe Kamiya of the University of Chicago and Barry Sterman of UCLA made seminal discoveries on the ability of an individual to control their own brain waves. Kamiya was focused on teaching humans to produce alpha waves spontaneously, and Sterman was training cats. Sterman was curious whether cats could voluntarily increase a specific brain wave in order to get a reward. His target brain wave came out of the sensorimotor cortex (SMR)—a region of the neocortex near the crown of the head. Of all the brain waves, it is most similar to the beta wave, since it cycles 12 to 15 times per second. Whenever the cats spontaneously produced SMR, they received a reward; soon they learned to produce it on cue.

Around that same time, NASA asked Sterman to investigate whether a certain type of rocket fuel might cause seizures. Sterman found that the rocket fuel did cause seizures in cats. However, he forgot to separate the cats in his two studies. Some of the cats from the SMR control study ended up in his rocket fuel seizures study. When he realized his error, he was about to throw away this data when he noticed something odd: the SMR cats did not develop seizure activity as readily as the others. In fact, their ability to withstand seizure triggers was much higher.

The EEG training had, in some way, protected the cats from seizures. This finding launched a flurry of research: Could EEG training be used to treat seizure disorders generally? In the 1970s, Joel Lubar began controlled trials investigating whether SMR could be used to treat ADHD, too. Neurofeedback as a nondrug therapy was born.

Neurofeedback's Many Applications

Neurofeedback has been applied to a dizzying number of disorders: autism, learning disorders, mood disorders, anxiety disorders, substance abuse, traumatic brain injury, and insomnia, to name a few.[5] In 2012, the American Academy of Pediatrics placed neurofeedback as a Level 1 or "best support" treatment option for ADHD, the same efficacy level reserved for medication.[6] In contrast to stimulant medication, whose effects tend to disappear as soon as patients stop taking them, the goal of neurofeedback for ADHD is to train weak attentional networks to function better on their own. Eventually, the attention networks strengthen to the point at which individuals can sustain attention—in many cases with less or even no stimulant medication.[7]

Neurofeedback protocols that train people to produce more beta and gamma waves have been found, in a few small studies, to lead to increases in IQ.[8] Delta and theta are high-amplitude, low-frequency (that is, slow) brain waves often associated with being asleep, but when present in large amounts while a person is awake, they tend to be associated with daydreaming and creative free association. Reducing theta has been shown to reduce daydreaming and distractibility.[9]

When I contacted local neurofeedback clinicians, case studies poured out. There was the boy who started out a problem child but transformed into a straight-A student, and the adult with lifelong fatigue and brain fog who finally energized after neurofeedback.

CAN NEUROFEEDBACK IMPROVE MEMORY AND LEARNING?

Neurofeedback has shown promising results for learning and athletic performance. Studies show medical students learning microsurgery faster after neurofeedback[10] and professional musicians developing expertise faster.[11] Olympic and professional athletes have cited neurofeedback for helping them gain an edge, including Olympic volleyball player Kerri Walsh-Jennings,[12] India's first individual Olympic gold medalist, Abhinav Bindra (who you read about earlier), and top pro golfer Bryson DeChambeau.[13]

CAN ATTENTION IMPROVE WITH JUST 10 MINUTES OF NEUROFEEDBACK PER DAY?

In an effort to provide a low-cost alternative to group mindfulness training, a Canadian research team recruited a group of 26 young men and women around 32 years old and randomly assigned each to either an experimental group or a control group.[14] The experimental group did just 10 minutes a day of neurofeedback-based meditation using a relatively inexpensive consumer device at home. The control group did 10 minutes a day of online math exercises. Both did their 10-minute intervention every day for six weeks. Initially, the neurofeedback group had higher motivation, but by the end both performed with the same accuracy on an attention task. However, the neurofeedback group performed their attention tasks faster. The neurofeedback group also exhibited greater well-being improvements, reduction in bodily symptoms, greater body awareness, and calm.

UPGRADING ATTENTION — AND ACCIDENTALLY BOOSTING IQ?

Through it all, I kept hearing about IQ gains. While I've never been a fan of using IQ as a measure, I was intrigued that clinicians were attempting to measure mental performance before and after interventions. Also, IQ had always been deemed a stable measure; the idea that a brief intervention might really change IQ...this was entirely unorthodox.[15] At a biofeedback conference in Oakland in 2014, I met Siegfried Othmer, a Cornell-trained physics PhD who switched his career from missile systems to neurofeedback after it transformed his ailing son's life.[16] I asked him about the IQ gains I'd heard about. It turned out that Othmer had published a case study about a pair of developmentally delayed twin girls who gained 22 and 23 IQ points respectively after neurofeedback—and maintained the gains at three retests over the following four years.[17] The paper included an excerpt from a letter that the twins' mother wrote after their treatment. "I can honestly say that it has made such a big difference in both girls. I don't know where to start!... The thing we are most pleased with is the progress they have made in school.... They reason and think when they talk. They've gone from being extremely dependent to being...the type that say, I'm going to go ride my bike, I'll check-in in a while." That paper mentioned six other papers showing neurofeedback's effects on IQ, with sample sizes ranging from 18 to 98 children. The IQ gains reported in the papers ranged from 4 points on the very low end to over 23 points. Most of the studies were conducted on children, so it's hard to know whether these kinds of gains would occur in adults, too, but the gains were large enough to catch my attention.

There are a few reasons I think they would. The fact that neurofeedback is so well proven at improving attention, the fact that attention is a gatekeeper to performance on IQ tests (after all, if you're not paying attention, you won't perform at your potential), and the fact that so many adults now struggle with attention—probably due at least partly to our smartphones—makes me think that neurofeedback has a good shot at improving adults' IQs, too. We'd need more research to see how widespread the effects would be, but if you're interested in this question for yourself, you could always take an IQ test before you begin training in neurofeedback, schedule a second IQ test for six months after the first test (as clinicians say this waiting period reduces the practice effects),[18] and train in neurofeedback in between. That way, you'd be able to see whether neurofeedback works for you.

LONG-TERM EFFECTS OF NEUROFEEDBACK

Although the clinicians I spoke with assured me that their clients' results last for years to come and that they only occasionally come in for "brightening" or "touch-ups," the research literature doesn't have much about the long-term effects of neurofeedback. To be fair, many interventions lack data on long-term effects, so neurofeedback is not alone in that. I did find a couple of studies in which researchers followed participants over time, though.

In 2013, there was a three-year follow-up of 25 healthy adults by researchers in Germany and the Netherlands.[19] The researchers compared the improvements of a group who were randomized to receive feedback from their own brain (true neurofeedback) against the improvements of a group randomized to get feedback from a prerecorded track of someone else's brain (sham neurofeedback). The researchers explored whether the neurofeedback group actually retained their brain-wave changes at the three-year follow-up. Remarkably, the biological changes remained. However, the researchers made an odd choice; they used a neurofeedback approach typically used on people with ADHD, reinforcing beta waves and decreasing theta waves. Apparently, this protocol was a poor fit, because the group did not measurably improve their cognitive performance.

MY NEUROFEEDBACK EXPERIMENTS TO IMPROVE FOCUS AND EXECUTIVE FUNCTION

I tried neurofeedback at home and in a clinic. For my at-home experience, I borrowed an EEG headset from a friend (who had ordered it online for around $250). A meditation app on my phone communicated via Bluetooth with the headset. The app told my headphones to relay the message of "good job, brain" by playing chirping bird noises. The message "try harder, brain" came via the whooshing sound of wind across sand.

The game-like basis of the neurofeedback mechanism appealed to the athlete in me. Every calm moment that the headset detected during a session earned me points in the app, which meant that by the end of each session, my goal was to beat last session's score. Since the signal between the headset and the app refreshed roughly every half second, over the course of 10 minutes I would receive around 1,200 opportunities to learn and update my brain state.[20] The relationship between my mental state and the resultant sounds from the app become an intensely personal competition. I found myself exploring, playing with, and toggling my mind in an entirely new way. I wondered whether the app would register a calm moment if I mentally counted backward from 100. What if I imagined a flame growing larger? If I focused on a point on the back of my head?

Within a few sessions, I was able to produce more bird chirps and less wind through the sand. My thoughts felt more defined, sharper, faster. Was neurofeedback measurably improving my cognition? It was time for another experiment.

Previous studies comparing neurofeedback to unassisted meditation had been inconclusive but promising.[21] I decided to compare the immediate effects of two meditation types: neurofeedback-assisted meditation using a consumer EEG headset and app, and a guided meditation using an app. My mind felt immensely clearer right after neurofeedback. It was much better than drinking tea or coffee; it was like getting a full night's sleep and then some. I felt somewhat clearer after the guided meditation, too. The biggest difference between the two was in the effect on my energy level; I felt much more tired after neurofeedback than I did after regular guided meditation. Going forward, I decided to keep using neurofeedback as a form of mental weight lifting—something that tired me out in the moment but would make me stronger over time. I would use guided meditation as a way to quickly clear my head after a stressful day.

After checking with the Biofeedback Certification International Alliance, I found clinics that offered neurofeedback sessions for $150 per hour. I ultimately picked Thomas Browne, a former national gold and silver medalist in track and field in Ireland who later trained and mentored under neurofeedback pioneer Joe Kamiya.[22] I ended up going for eight sessions over the course of a few months. Each session lasted about 30 minutes.

Dr. Browne asked for my primary complaints. I answered: brain fog, low mental energy, executive function issues (including distractibility and working memory), and some anxiety. He focused on increasing my beta and alpha waves, decreasing theta, and decreasing high beta (to reduce anxiety). After the electrodes were attached to my scalp and sending signals of my brain waves to the desktop computer, Dr. Browne placed headphones over my ears. I closed my eyes, sat back in the office chair, and listened to the feedback music being created in response to my brain-wave patterns. As with my at-home setup, whenever I got into the target brain-wave pattern—my brain producing closer to a normal ratio of high- and low-frequency waves with more beta and alpha waves, fewer theta and high beta waves—the sounds coming through the headphones changed to the reward pattern. Whenever I fell out of the optimal ratio, the gentle rebuke of less pleasant sounds came through the headphones instead.

After the first session, I felt surreally better for at least an hour. I felt like I was bathed in warm, golden light—as if my whole body and brain were glowing. I felt giddy. My brain felt different, cleaner. Before the session, I'd been running late, and my thoughts had felt like old grocery bags tossed around by the wind. Now, my thoughts came crisply, in an organized and prioritized order. After a few sessions, my mind felt like it had come back from a Caribbean vacation. There was no struggling for words when I spoke, and my speaking felt fluid.

In terms of measurable results, my brain waves had altered significantly. For my first session, the computer screen displayed charts comparing the ratio of my beta waves to theta waves. Dr. Browne remarked that they looked as out of whack as those of kids with ADHD diagnoses; the low-frequency brain waves that dominate in sleep were prevalent even though I was awake. Within a few sessions, however, mine were back in the normal ratio range. The changes stuck, too; at the start of each subsequent session, the ratios looked as they had at the end of the previous session.

My cognitive test scores improved the most in my personal weaknesses: visual perception and executive function, including working memory and certain types of short-term memory. My visual perception improved by about 30 percent, and executive function, my second-weakest area before the neurofeedback treatment and the one I cared about most, improved by over 20 percent. Surprisingly, one of my strongest areas improved, too: I improved by 16 percent in verbal short-term learning ability.

Comparing at-home to in-clinic neurofeedback is tricky. At-home neurofeedback is much more convenient, you can go for as long or short as you want, and there's no commute. As we know from the discussion of placebos in chapter 13, though, one benefit of an in-office intervention is the very formality of it—and the feeling of being cared for that comes from a person administering an intervention to you. There are other comparison problems, too. Since the in-clinic treatment occurred after I'd already experienced gains from the at-home combination treatment, maybe the at-home treatment stole all the low-hanging optimizations first, leaving the harder problems for the in-clinic treatment.

The at-home treatment and the in-clinic versions each boosted my executive function by at least 20 percent. The at-home was a combined treatment of neurofeedback and medication, whereas the in-clinic was pure neurofeedback. Maybe the combination was better? I predict that remote therapy over the internet will become more popular. Already, a few companies offer it—including an Israeli startup called Myndlift that allows people with inexpensive but high-quality neurofeedback headsets (like the Muse EEG headband) to get treated by experienced professionals either locally or over the internet. Hopefully, the convenience of at-home neurofeedback will become more commonplace.

TAKEAWAYS

1. Neurofeedback is a technique that allows you to learn to control your own brain waves.
2. Certain protocols have helped with executive function, others with learning and memory, as well as ones that have helped surgeons and musicians develop expertise, and athletes to up their game.
3. Neurofeedback has minimal side effects. Those that do exist include fatigue, headaches, or nausea.
4. The enhancement effects seem to last, at least for improving executive function. There are many reports of sustained IQ gains, too.

Chapter 17

Serious Games

"The game of Chess is not merely an idle amusement.
Several very valuable qualities of the mind, useful in the course
of human life, are to be acquired or strengthened by it, so as to
become habits, ready on all occasions."
— Benjamin Franklin, "The Morals of Chess"[1]

> **Time Investment:** 38 minutes
> **Goal:** To understand the power of games to upgrade
> your mental performance and learn how to
> incorporate them into a neurohacking experiment

As I walked up the stairs, a telltale tune floated down. The oboe, clarinet, piano, harp, strings, viola, cello, and percussion told me all I needed to know: my husband, Varun, was at it again. Playing the video game *Zelda: Breath of the Wild* had become his end-of-workday ritual. He had first started playing it on the flight back from a conference. The pattern had become clear: when he'd had a stressful day, he craved the video game. And the amount of time he spent playing seemed to be proportional to how stressful his day had been. Initially, I pushed back; why didn't he want to talk instead? But there was no denying the happiness that the game brought him. Over the course of a session, his body language shifted from silently grumpy to focused and intent. Eventually, he looked downright relaxed, even cheerful. What was this black magic?!

I didn't grow up playing video games, but I learned to respect them. After observing how so many of my highest-achieving classmates at MIT played video games regularly, I grew skeptical of the "video games rot your brain" rant. There had to be more to it than that. By midway through my time at MIT, I had become so intrigued by games that I'd even started building educational video games of my own in a research group at the MIT Media Lab. Now, with the obvious mood boost Varun's daily gaming provided him I found myself wondering: Could video games be used as a mental upgrade?

What Is a Game?

The goal for the video games I built at the MIT Media Lab was for the player to have fun *while* they learned. Not some chocolate-covered broccoli trick, but real learning, real fun, mixed equally. The game I built was called Physical Physics. I put a weight sensor in the player's shoe that communicated wirelessly with a video game. Every time the player moved, they would see their movement displayed on a chart. By moving and watching changes in the chart, the player developed a more personal, intuitive understanding of physical concepts like displacement, velocity, and acceleration. Later, I learned that there is a name for the type of game I had built: it is called a "serious game."

In short, serious games are games played with a serious purpose. They are designed to teach the player something, to change their real-life behavior, or to accomplish some other "serious" goal. Serious games aren't just reserved for the traditional educational setting; they crop up in corporate, military, and healthcare trainings, too.[2]

Unlike in many traditional educational settings or medical interventions — where we don't always want to do what we're told — games motivate us. As we discussed in Part II, psychological pioneer Mihaly Csikszentmihalyi coined the term "flow" to describe that blessed mental state some people call "being in the zone." Csikszentmihalyi noticed that when people were faced with a task that was too difficult for their current skill level, they became anxious. Conversely, when the task was too easy for them, they became bored. When the task was just right for their skill level, however, it often led to flow.[3] Matching your skill level to in-game tasks set at just the right level makes for a pleasingly addictive game and for flow.

Levels of Skill and Difficulty in Flow Versus in Boredom or Anxiety[4]

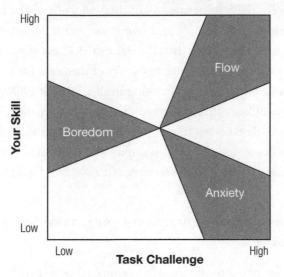

Games' addictiveness lie in the rewards: unlike most of the rest of life, a game provides you with nearly instant feedback. Our brains are wired to crave feedback, and when we get it, we seek more of it.[5] This helps our brains learn new information or habits quickly.

Can Games Change the Brain?

When I started looking into data on gaming, I was surprised to find out that one of the pioneers in this space is...the US Navy's Office of Naval Research (ONR). Why would the military want their staff to play video games?

In a 2010 interview on Pentagon Web Radio's "Armed with Science" audio webcast, Ray Perez, a program officer at the ONR's Department of Warfighter Performance, explained that they had found 10 to 20 percent higher performance levels in certain cognitive abilities among video gamers.[6]

While it's intriguing to note that players of certain video games exhibit cognitive advantages, we can't just assume that the video games *caused* the cognitive improvements. Maybe the people who play video games just started off sharper to begin with. To answer the question of whether video games improve cognitive performance, we need experiments. Many of the secrets we've uncovered about what gives rise to neuroplasticity are present in well-designed games. There is, in fact, mounting evidence of the structural and functional changes that games can cause in the brain. Structurally they can include large-scale changes, such as the size or thickness of parts of the brain. Functionally they can affect which networks of the brain turn on or off when you perform an activity. In brain-imaging studies involving 800 video-gaming participants, neuroscientists "observed changes in virtually all parts of the brain, such as in fibers connecting to the visual, temporal and prefrontal cortices, the corpus callosum, the hippocampus, the thalamus, association fibers like the external capsule, and fibers connecting the basal ganglia."[7]

STRUCTURAL CHANGES: WHICH BRAIN AREAS TEND TO CHANGE FROM GAMEPLAY?

In the article describing the findings involving 800 video gamers, the neuroscientists "found relevant changes in prefrontal regions." These included areas you may recall from our discussion of executive function and learning and memory. Prepare for an eyeful of neuroanatomy terms: the dorsolateral prefrontal cortex (dlPFC) and surrounding areas, superior and posterior parietal regions, the anterior cingulate cortex (ACC), the cerebellum, the insula, and subcortical nuclei, as well as the striatum and the hippocampus. Regions that researchers have found to be active during video game play include ones that have to do with physical movement, such as the motor and premotor cortex. Other regions include those having to do with integrating information from your five senses and processing of visual attention, such as the top and bottom of the parietal lobe. Activity changes that relate to learning, emotion processing, and memory occur in areas such as the amygdala and parts of the entorhinal cortex.

FUNCTIONAL CHANGES: WHICH COGNITIVE FUNCTIONS TEND TO CHANGE FROM PLAYING WHICH KIND OF GAMES?

The review above looked at functional differences in the brain related to video games across 100 studies that included 3,229 participants. For instance, when gamers played the puzzle game Tetris they ended up with, on average, a thicker cortex and greater brain efficiency.[8]

Research into entertainment-focused games is still a new field, but some studies are starting to shed light on which types may be more promising for cognitive enhancement than others. In 2014, Florida State researchers randomly assigned 77 undergraduates who were not frequent video game players to play either Portal 2, a puzzle-platform game, or Lumosity, a brain game, for eight hours over the course of one to two weeks, to test the effects on their problem-solving ability, spatial skills, and persistence.[9] I was especially curious about the outcome of this study, as I used to work on the research team at Lumosity. The amount of training time allocated in this study was much less than what other brain-training studies have found to be effective, such as the 10 hours used in the prestigious ACTIVE study, which you'll learn about later in this chapter. Across all three measures (problem-solving ability, spatial skills, and persistence), the differences between the pre- and post-game tests were negligible for the Lumosity group, but, interestingly, the Portal 2 group actually improved on all three, even in that short time frame.

In another 2014 study, researchers at George Mason University compared the effects of an entertainment-focused game and two different brain games on participants' working memory and problem-solving. They recruited 42 healthy older adults and randomly assigned them to six weeks of training with either Rise of Nations (a real-time strategy game they thought might enhance strategic reasoning), Space Fortress (a brain game designed by neuroscientists to enhance visuomotor/working memory), or Brain Fitness (a brain game suite designed by neuroscientists to enhance auditory perception).[10] The result? In terms of working memory, the group that improved the most was the group that played Space Fortress. On a measure of problem-solving, the group that improved the most was the group that played Brain Fitness. Rise of Nations (the entertainment-first game) came in last. Does that mean that Rise of Nations is a waste of time? Not necessarily, but if your goal is to improve your working memory or problem-solving ability, perhaps Space Fortress or Brain Fitness should be higher on your list of interventions to try.

Brain Games to Upgrade Executive Function in Older Adults

The kinds of treatments that work well for older adults may not work for younger adults, and vice versa, so read this section if you're especially interested in upgrading older brains. Older adults have long been encouraged to do daily puzzles and crosswords to maintain mental agility. Are there any more targeted games aimed at offsetting cognitive aging that have real science behind them? While the brain game industry has not always prioritized good science, one game maker has consistently bucked that trend. The next two studies we will discuss involve the evaluation of brain games made by neuroplasticity pioneer Michael Merzenich, an emeritus professor at the University of California–San Francisco and a member of both the National Academy of Sciences and the National Academy of Medicine.

In 2013, the *Journal of Aging and Health* published findings from the ACTIVE study (which stands for Advanced Cognitive Training for Independent and Vital Elderly and was an NIH-funded, randomized controlled trial involving nearly 3,000 participants 65 to 90 years old).[11] In contrast to the control groups, the brain game players enjoyed improvements in their memory, reasoning ability, and visual-processing speed. In addition, they reported that they had improved in their independent activities of daily living (dressing, cleaning and managing a house, managing money, shopping, taking medications, and so on)[12] and had fewer at-fault automobile crashes (as recorded by state departments of motor vehicles) than the control group. At two-year, five-year, and even ten-year follow-ups, although no longer playing the brain games, the participants' gains from long-ago gameplay persisted.[13]

Another large-scale randomized controlled trial, this one with 620 middle-aged and older adults, was called the IHAMS, short for Iowa Healthy and Active Minds Study. It used the same brain games but used computerized crossword puzzles in the control group.[14] The IHAMS showed that the same brain games tested in the ACTIVE study worked better than crossword puzzles. Furthermore, it showed that middle-aged adults could benefit from the same games that older adults had. Because some participants who were randomly assigned to play more than others also enjoyed higher gains, the study showed that gameplay benefits were dose-dependent; more was better. Finally, like the ACTIVE study, the gains lasted far after the gameplay had ended. At a one-year follow-up, they had retained their gains, even though they were no longer playing the games.

Other brain games have bested crosswords in experiments as well; researchers from Lumosity, Wheaton College, and Wayne State University randomly assigned nearly 5,000 online participants to either crossword puzzles or Lumosity games. In 2015, they reported that the brain games group enjoyed greater gains on most untrained cognitive measures, as well as higher self-reports of improvement in their daily lives.[15] Unlike the ACTIVE or the IHAMS, however, the Lumosity study was coauthored by employees of Lumosity, thus there was more potential for a conflict of interest.

In 2013, a group at the University of California–San Francisco developed a 3D racing game called NeuroRacer.[16] They recruited 60- to 85-year-olds and compared their multitasking skills to those of a group of 20-year-olds. After the older adults had played the game for 12 hours total (1 hour per day, 3 days per week, 4 weeks), they multitasked better than untrained 20-year-olds. There were significant attention-related changes in their brain waves as well, as measured by EEG. Impressively, the performance gains persisted at a six-month follow-up. Not surprisingly, this research made the cover of the scientific journal *Nature*.

Do Video Games Cause Violence or Addiction?

Some worry that video games encourage violence. Scientifically, the evidence is mixed. A German study published in *Molecular Psychiatry* in 2018 found that in adult participants with no known psychological disorders, after about 30 minutes of playing a violent video game (Grand Theft Auto V) daily for 8 weeks, game players reported no more aggressive or violent tendencies than participants who had been randomly assigned to play Sims 3 for the same time period.[17] These findings do not support the idea that violence in video games promote violent behavior in life. However, a review of 24 studies focused on the relationship between video games and physical aggression in the real world took a strong stance against violent video games based on their findings.[18] The researchers found that across these studies, which included more than 17,000 children ranging from 9 to 19 years old, violent video gaming was associated with significantly greater levels of overt physical aggression (hitting family members, trips to the principal, and so on) in these children's daily lives. These findings support a link between violent video games and violent behavior, particularly in children and adolescents.

There is also concern about video game addiction. Recently, the World Health Organization added "internet gaming disorder" to the edition of its International Classification of Diseases published in 2019,[19] and the

American Psychiatric Association (APA) included it as a potential mental disorder in the 2013 publication of the *DSM-5*.[20] While many adults play video games, estimates cited by the APA suggest that less than 1 percent of the general population meet the clinical diagnostic criteria for video game addiction, and other preliminary research suggests that video game addiction tends to be comorbid with other mental health challenges.[21] It is important to note that, as with all clinical diagnostic criteria, for a person to meet the criteria for video game addition, their relationship to gaming must cause "significant impairment or distress" in other aspects of their life.[22] So when we're talking about clinical addiction, we're not just talking about someone who enjoys playing games, but someone who is finding that their relationship to gaming is making it difficult to take care of themselves or fulfill their other life duties, like school, work, or relationships. Thus, while many people can enjoy gaming as a healthy, balanced part of their life, others might struggle with addiction or feel like they are losing themselves to gaming.[23] Like any addiction, video game addiction can wreak havoc on people's personal and professional lives, even tearing apart families. If you have a problem, there are resources at gamequitters.com.

Can Games Improve Executive Function?

Specific types of video games result in physical changes to the brain and measurable improvements in executive function. As we discussed above, experiments in which nongamers start playing video games reveal that gameplay leads to thickening in areas of the cortex and increased efficiency in certain brain networks. Best of all, there seems to be transfer to nongame performance—that is, gamers improve their performance on seemingly unrelated cognitive tests.

FOR SELECTIVE ATTENTION, THE ACTION IS IN ACTION GAMES

If you've ever been in a room filled with loud, bright distractions and struggled to focus on just one thing—say, what the person next to you was saying—you've experienced the limits of your "selective attention." This type of attention comes up in many circumstances in life. When driving, we need to filter out all the irrelevant visual information coming at us and pay attention to the traffic signal up ahead. When taking a test, we need to drown out the buzz of the lights, the click of a classmate nervously tapping their pencil, the overwhelming perfume that the test proctor is wearing. When it comes to improving selective attention, the fast-paced, demanding nature of action video games seems to help more than other types of games. In a 2017 systematic review conducted by researchers at Spanish and American institutions,[24] the authors found that action video games improved selective attention better than did slower-paced video games, such as role-playing games,[25] puzzle games,[26] or strategy video games.[27] The researchers believed this was due to action video games requiring a lot of attention under timed conditions.

BRAIN GAMES THAT COULD MAKE YOU SMARTER

There has been a lot of controversy around commercial brain games. Some of this is simply to do with how young the field is, and some of it has to do with how much people want brain games to work. We know plasticity is possible, and we just keep hoping that something cheap, noninvasive, and fun will be the path to unlocking a brain upgrade. To be honest, I'm not sure that brain games designed by neuroscientists offer the most effective interventions in this book. Given how accessible they are, however, they should still be considered.

All of the studies I mention, both above and below, were published in respected, peer-reviewed journals. Some were conducted on thousands of participants. Also, I've chosen studies in which the effect of a brain game was compared to some other interesting activity—say, a crossword puzzle or an entertaining educational program. (Often, brain games research compares the effects of a brain game to doing nothing at all, which is not a very high bar.) I also mention studies whose participants actually transferred the skills they learned in the brain game to untrained tasks in the real world, and other studies where the brain game group enjoyed fairly long-term changes in their mental performance.

While I tried to include high-quality studies, these are still far from perfect. Most of them were conducted closely with the makers of the games themselves (or the task was later made into a commercial game), so there is always a question of bias creeping into the results or the design. And, of course, you'll only ever know whether something will work for you if you try it out on yourself. That said, let's dive in.

The first type of brain game with scientific evidence indicating it could improve executive function is called the "dual n-back." You have to track visual information and auditory information. There are different designs, but you're typically tracking two streams of information, such as listening to a string of letters being spoken and watching a series of flashes on a geometric grid. If you are playing the "2-back" version of the game, you press a button as soon as you hear the same letter as the one you heard two presentations back. You press a different button if you see the same tile light up that lit up two presentations ago. If you're playing the "3-back" version of the game, it would be the same, except you'd be watching for the same tile lit up from three presentations ago and listening for the same letter as the one spoken three presentations ago. The core challenge is to track and act upon both streams of information at the same time. Transforming the n-back from a test used by cognitive scientists to *assess* a person's working memory into a game used to *improve* working memory was part of the innovation. The type aimed at improvement actually gets harder as you do it—it adapts to your level.

Using the dual n-back for cognitive enhancement remains controversial in many parts of the scientific community, because scientists struggled to replicate the promising results found in an initial study. The initial researchers— most notably, Susanne Jaeggi and her colleagues at the University of Michigan— found that their young, healthy, adult participants improved significantly in untrained areas of cognition after playing dual n-back games.[28] The researchers boldly stated that the participants had improved in something called fluid intelligence. Fluid intelligence relates to comprehension, solving novel problems, and learning new information. It is described in contrast to crystallized intelligence—the ability to manipulate knowledge and expertise that you already possess, such as facts. Jaeggi's work attracted a lot of attention—and criticism. At this point, there are some promising meta-analyses and small-scale studies that came to positive conclusions, but I did not find any large scale, multi-site studies comparing the effects of dual n-back to some other comparable brain-training intervention.

Anecdotally, some swear by these games and believe them to be responsible for jumps in their overall IQ. For instance, veteran biohacker and

Bulletproof Coffee founder Dave Asprey swears that he gained 2.75 IQ points for every hour that he trained on the dual n-back.[29] For people who struggle with working memory and other aspects of executive function—such as children and adults with ADHD—these kinds of games could be especially worth considering.

If you are at all curious, the barrier to trying these games is low. There are free dual n-back games you can try at home, and there are fancier and more personalized for-pay versions that come with money-back guarantees (for example, if you don't gain 10 to 20 IQ points on an official IQ test after trying neuroscientist Mark Ashton Smith's version of a dual n-back, he will refund your payment).[30] I've played a few versions of various types of dual n-back over the years casually. All I can say for sure is that the games are very mentally challenging. As always, individual differences rule; your mental mileage will vary with this intervention as compared to anyone else. If you're curious, I don't see any harm in giving it a try.

Can Games Improve Emotional Self-Regulation?

Researchers posit that there are several main ways in which games improve our emotional self-regulation.[31]

GAMES CAN CHANGE YOUR EMOTIONS DIRECTLY.

By distracting you from your previous thoughts, games can sometimes replace whatever you were feeling before with the feelings brought on by the game. How might this distraction or emotion replacement help you gain better control over your emotions? The power comes in knowing that this happens and using it as a tool intentionally. I asked my husband, Varun, why he had gotten in the routine of playing games during an especially stressful period at work. He explained that he used the game as first aid for his stress. He also knew that this particular problem was short-term—he just needed to get through the rough patch at work, but it would be over soon. In the meantime, he would intentionally use the game's distraction and artificial mood lift as a short-term coping mechanism.

Had anyone else noticed this emotionally regulating superpower of games? In 2018, researchers from Oxford, Cambridge, and the Karolinska

Institute in Sweden enrolled emergency room patients in an experiment. These patients had ended up in the ER because they had been involved in a traffic accident. Each patient was told that the study's goal was "to examine how simple activities affect flashbacks and other symptoms." Then, each person was randomly assigned to either play a 20-minute game of Tetris or do a 20-minute "activity log," a recollection of the tests they had taken and other details of their time in the hospital thus far. They noted things like "talked to family," "scan of chest and abdomen," "nurse sorted out drip." A week later, the Tetris players had far fewer flashbacks than the activity loggers.[32] Was it simply that they'd had a welcome distraction during a stressful time—as Varun had with his video game play? Not exactly. In fact, the researchers noticed something far more specific. They believed that the brain's visual system was so engaged by Tetris's high demands that it was not able to consolidate the visual memories of the traumatic accident that had just occurred. By interfering with the ability to process and store those visual memories in the usual way, Tetris had blocked the survivors' ability to fully store the memory of their traumatic event. Robbed of a full memory consolidation, the brain couldn't produce PTSD flashbacks later on as easily, either. What this study shows us is that the right video game played at the right time can do more than just distract—it can alter how we process the emotional event itself, making the later regulation of our emotions that much easier.

GAMES CAN TRAIN YOU IN THE "ART OF FAILURE."[33]

Many coaches assert that games provide opportunities for kids to experience success and failure and learn the most important lesson of all: that both situations are temporary, and it's what we learn from them that provides the greatest value long-term. On one occasion, I remember, I lost a match to another squash player whom I had beaten handily in the first game. What I discovered over the course of the next few games was that her early loss had triggered a moment of self-reflection. Rather than being crushed by her initial defeat, she took the 90-second break between games to analyze her past actions and mine. I overheard her talking with her coach and her parents about how to approach the situation differently than she had. After the break, she was armed with a new plan. During my 90-second break, I basically just hung out on the side of the court, smiling, drinking water, and feeling good about my assumed win. I may have won that first game, but my opponent went on to win the match.

GAMES CAN HELP YOU REGULATE EMOTIONS BY ENGAGING YOUR IMAGINATION.[34]

Video games often involve picking an avatar, a name, and playing the same scene multiple times or as different characters in order to uncover hidden truths and rewards. These types of make-believe provide a valuable real-life skill: the ability to imagine different versions of yourself and of your life. Part of emotional self-regulation is about changing how you see or feel about a situation, and that can lead you to change how you behave, ultimately empowering you to change yourself or your life. All of this is hard work, but it relies on a critical first step: imagining that things *could* be different. It's about imagining that you *could* have felt differently, that you or someone else *could* have behaved differently. Playing games builds that critical muscle of imagination.

GAMES TEACH YOU EMOTIONAL CONTROL DIRECTLY.

What if a game got harder the more anxious you got? A game called Mightier does exactly that for children. Using biofeedback, the game increases the difficulty level as the player's heart rate increases. It is designed to help children who struggle with self-control to gain awareness of their emotions and learn to control them. Children learn to associate the increased challenge with a lack of emotional self-control. This cues them to stop, take deep breaths, and resume the game when they are ready. The game lowers the challenge level as soon as they regain calm. After 12 weeks, researchers at Boston Children's Hospital and Harvard Medical School found that the children who played Mightier had improvements in their real lives, too; they had 62 percent fewer outbursts and experienced 40 percent fewer incidents of oppositional behavior, and their parents reported being 19 percent less stressed.[35]

While it has been studied less rigorously than Mightier, there exists a biofeedback-based game that targets adults: Nevermind. It was developed by students at the University of Southern California. In it you learn to control your heart rate in the context of a horror-suspense world, otherwise the world becomes scarier. It is not intended to treat a clinical disorder, but in a small study some participants did show an increased ability to self-regulate their fight-or-flight response while playing it.[36]

If you are wondering whether there are games designed to help you exert self-control across your whole life, the answer is yes. In her popular TED Talk,[37] video game designer Jane McGonigal shared a harrowing story: her own battle with suicidal thoughts. McGonigal had suffered a concussion that left her in constant pain. As her recovery dragged on and on with little obvious improvement, she began to lose hope that she would ever recover her brain's previous levels of function. McGonigal eventually saw a unique opportunity: she would use her game designer skills to motivate her own concussion recovery. By converting the everyday tasks required for her concussion recovery into a game, she slowly began to regain control of her life. As McGonigal finally recovered, she began to wonder: Could her game— that she had, by now, dubbed SuperBetter—help others, too?

As it turned out, the answer was yes, to the tune of millions of people worldwide. SuperBetter attracted people fighting chronic diseases, recovering from traumatic injuries, and struggling with depression after sudden misfortunes. It also appealed to people just looking for a more creative way to tackle boring, everyday responsibilities. When SuperBetter later attracted the attention of other researchers, McGonigal was thrilled with the results. A group of 283 people struggling with significant depression symptoms were randomly assigned to either play SuperBetter for 10 minutes a day for one month or to be put on a waitlist to later get to play the game. At the end of the month, the depressive symptoms of those who had played SuperBetter had improved on average by over 23 percent relative to those who were on the waitlist.[38]

While SuperBetter uses classic game elements like power-ups, quests, and allies to motivate you to play the game more, I found myself wondering whether any self-regulation games existed that use real-world rewards. If you struggle to get yourself to exercise, one of these apps may be worth testing out. A 2019 study of a game-like app called Sweatcoin showed that, over the six-month period following download, users increased physical activity by around 19 percent.[39] As you exercise in the real world, you earn in-app currency that allows you to buy items in the real world—or donate to real organizations. The currency conversion is made possible through partnerships with brands that want to connect with health-conscious consumers, as well as governments and insurers that want to reduce healthcare costs. Related apps include Achievement, Dietbet, and Charity Miles.[40]

Can Games Improve Memory and Learning?

If ever there was a neurohack, this is it. This one is a biggie. What if you could spend an hour studying for a test and perform as well or better than classmates who spent months studying? This may sound like the fever dream of every panicked student before finals period, but it was the reality for a lucky set of British biology students in 2013.[41] In the words of one student from the class, "We only had a one-hour Spaced Learning review session... Most of us did better on the exams after [this], even though we did no studying at all. I went from an A, B, and C to straight As and an A+. It was amazing."

Depending on the version of spaced learning implemented, the results can be 200 percent better[42] than conventional methods, or even more.[43]

The Forgetting Curve[44]

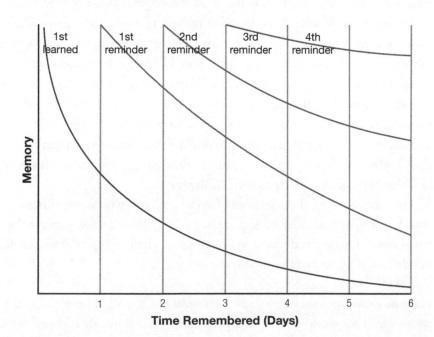

1st learned · 1st reminder · 2nd reminder · 3rd reminder · 4th reminder

Memory

Time Remembered (Days)

When I came across these studies, I breathed an audible sigh of relief. *This* was the kind of improvement I was searching for when I first started out on my own neurohacking journey. Confession: spaced repetition didn't quite belong in any of the chapters, though. It's not exactly a game, but it is a phenomenal intervention, no matter what you call it.

What is spaced learning? If you remember back to the successful neurohackers we met at the beginning of the book, there was a trick that the *Jeopardy!* winner used, a software program. What was the secret sauce in the software? It goes by many names — spaced repetition, spaced retrieval, the spacing effect, spaced learning. The concept is simple: You enter information to be remembered into flash cards. Then, you get a reminder to review that information *just* before you would have forgotten it. While researchers have tried many versions of the reminder schedules, the basic idea came from studying the average timing of human forgetting.

In 2011, researchers from Purdue University compared the effects of traditional study methods with those of spaced retrieval.[45] While 75 percent of the student participants believed that the traditional methods would lead to better scores on the test, 84 percent of them performed better using spaced retrieval. Furthermore, those who benefited from the nontraditional technique benefited *much* more: they were better by between 36 and 43 percent, on average.

If you're out of school and wondering how to apply this, consider any learning goals you have. If you can turn your learning into flash cards, you can benefit from spaced learning. That goes for foreign languages, new terminology if you're in a fast-changing industry, the names and faces of new colleagues at work … the applications for spaced learning are limited only by your desire to learn. If you don't want to have to create flash cards from scratch, there are many education apps available that use spaced repetition as their engine to help you learn — language-learning apps are particularly fond of including spaced learning as their "killer feature."

Here are some recommendations for those of you who want to improve your learning in particular subject areas. I've highlighted a few games below that I've tried myself and found to be engaging, challenging, and accessible, but this list is by no means exhaustive.

Communication: I wanted a game that would help me hone my communication skills for work, school, and personal relationships, so I looked for a mobile app to fit the bill. Ultimately, the games and exercises in Elevate felt the most engaging, challenging, and fun. The beauty of educational games on the phone is that the sessions are bite-size. I typically played during Lyft rides or other waiting periods sprinkled throughout my day—any time I had a few spare minutes. The company published positive research on its own app in 2015 using tests they developed themselves—I would trust it more had it come from a third-party research team with widely used tests.[46] A group of 146 participants were randomly assigned to play the games and 125 were randomly assigned not to play the games. Both groups took pre-tests and post-tests. The group assigned to play Elevate five times a week scored nearly 70 percent higher than those not assigned to play. Also, as with most brain training, having more opportunities to practice helped more; the people who played the most (more than four times a week) scored 17.5 percent higher than those who practiced less than two times a week and 9.5 percent better than those who practiced two or three times a week. This matched my experience playing it; casually, I observed that on days that I played Elevate, I tended to have fewer "brain farts"—moments when I forgot a word or lost a train of thought. This type of observation is problematic, though, because the days I didn't play Elevate may have simply been days I was busier, more stressed, and less well rested—exactly the type of day you would expect to have more "brain farts" anyway. Overall, though, I felt engaged and challenged by the game.

Foreign languages: As with the quest to hone my English communication in just a few minutes a day, I began wondering if I could improve my Spanish in an ad hoc way, too. Duolingo, a free and popular language app, had fun, short lessons that involved a lot of translating from Spanish to English and English to Spanish. This meant that I wasn't fully thinking in Spanish, but the fun, retrieval-focused approach made me feel a sense of daily progress. Professor Roumen Vesselinov at Queens College found that Duolingo taught the equivalent of an introductory college course in 35 hours of practice,[47] and the paid language program Rosetta Stone took about 55 hours to get through a college semester's worth of material.[48] If you're working on grammar and vocabulary, perhaps Duolingo would be the more efficient app to try, but for a more immersive experience that challenges you to think in the language, Rosetta Stone might be the better bet even though it will take up more of your time.

Sports: IntelliGym's "low-fidelity" versions of sports such as basketball, hockey, and soccer in video game form have been adopted by professional teams. Why not just play the game itself—why play a video game version of a game? Some organizations believe that giving players a bare-bones version of the game to play on a screen, without the distractions of physical fitness or the screams of fans, can help athletes improve their mental models of the sport. Furthermore, more practice time—even when players are sitting in a chair— might provide the edge that brings their team a big win. The USA Hockey National Team Development Program averaged 42 percent more points per hockey game[49] after its team started training with IntelliGym. In addition, they found a reduction of 15 percent in on-ice injuries and a 28 percent reduction in head injuries.[50] This speaks to the potential of these games to train in-game intelligence[51] (although, as always, research sponsored by the maker of a product is never as convincing as third-party research).

Math: Khan Academy, with more than 70 million users as of 2018, is a website with free instructional videos on a range of topics. It's especially beloved by teachers, parents, and students for its math videos. While I've not seen randomized controlled trials showing that Khan Academy is superior to other math instruction methods, one group of researchers at Georgia Institute of Technology showed how Khan Academy incorporates game principles.[52] Perhaps gamification is part of its secret success sauce.

Speed reading: There's mixed evidence about whether speed reading is actually a good idea. There is a major concern that as you increase your speed, comprehension falls.[53] Speed reading might be a bad idea especially for dense or technical text. Still, if you want to learn skimming and speed reading, there are plenty of apps that gamify the process like Spreeder and ReadMe.[54]

Can Games Improve Creativity?

To be honest, I was a little disappointed when I looked through the evidence on video games' effect on creativity. At this point, the research just isn't clear yet—probably because both fields are relatively new. I struggled to find conclusive scientific studies. Still, a few jumped out.

Researchers at Penn State and Sungkyunkwan University in South Korea performed an experiment in which they used video games to alter participants' moods and then tested their creativity to see which moods provoked the most creative results. They altered the subjects' levels of emotional intensity by having them play a low-, medium-, or high-intensity game of Dance Dance Revolution (a rhythm and dancing game that involves stepping on targets on a mat in time with the music). They found that the optimal mood for creativity seems to be either when people were feeling negative but calm, or when they were feeling strongly positive. To get yourself in this state, you could replicate the study by either playing Dance Dance Revolution for five minutes (they had subjects dance to the four-minute song "Pump Up the Volume" by MARRS). Oddly enough, the other combination that led to enhanced creativity was when participants played Dance Dance Revolution at a low-intensity level and were then induced into a bad mood by being told they had incorrectly identified expressions on people's faces in an array of photos.[55]

In another study about games and creativity,[56] 352 participants were randomly assigned to one of four groups: One group was told to play Minecraft without any instructions. Minecraft is a sandbox-style video game that allows the player to explore an open world and create anything they want with the tools provided. A second group was told to play Minecraft but was urged to "be as creative as you can," a third group was told to play a NASCAR racing game, and the fourth watched a TV show. Then all the participants were asked to imagine an alien who grew up on a planet different from ours. If the participant drew it as very different from a human—say, the creature was asymmetrical, had only one eye, had more than four limbs—it was deemed more creative. The result of the study was that the group with minimal instructions, just told to go play Minecraft, actually produced the most creative alien drawings. All of the others, including those who were explicitly told to "be creative," were actually less creative. The researchers speculated that perhaps people perform worse when they are being extrinsically incentivized than if they're given no instructions. In the latter situation, they can self-motivate and take intrinsic joy in the task.

All of this shows us that games are powerful, and they can be used in ways that are negative. Choose your games wisely. In the next section of the book, you'll encounter a set of interventions that are sexy, a bit complex, and possibly a bit riskier than what you've seen so far.

TAKEAWAYS

1. Games have been shown to help with executive function, memory, and emotional regulation (including relieving anxiety and depression).
2. For older adults, brain games can help combat aging's effects on memory and executive function.
3. Certain games may help with creativity, such as Dance Dance Revolution and Minecraft.
4. All games are not created equal; choose the type of game based on whichever mental target you are trying to upgrade. To improve your odds, pick a game that performed well in scientific studies.

PART IV

Select an (Advanced) Intervention

Chapter 18

Zapping for the Better

Time Investment: 30 minutes
Goal: To understand the power of neuroelectrical stimulation to upgrade your mental performance and learn how to incorporate it into a neurohacking experiment

When your lids are drooping, you're fighting a deadline, and that last cup of coffee did nothing, you might consider some wild things to stay awake. But you'd probably draw the line at, say, sending an electrical current through your head. What if it induced savant-like problem-solving abilities, though? What if it caused you to play music as if you had jumped straight to a professional level? What if it caused you to suddenly think of new idea after new idea?

Still...if you ever saw or read *One Flew over the Cuckoo's Nest*, you probably recall the patients' memory loss and glazed stares after electroconvulsive (shock) therapy. And didn't Dr. Frankenstein use electricity to create his monster?

During the summer between high school and college, my dad showed me an article that, shall we say, sparked my interest (sorry, couldn't resist!). It was in the *New York Times* magazine and titled "Savant for a Day." The nervous journalist sat in a University of Sydney basement as an unnervingly gleeful professor turned on his "creativity cap." Under the electromagnetic coils, the journalist's artistic ability shot up: He went from stick figures to acceptably drawn felines in less than 10 minutes. Fulfilling the dreams of copyeditors everywhere, he also went from repeatedly missing the extra "the" in a sentence to noticing the error instantly. I decided that when I got to college that fall, I would take on this brain stimulation stuff. Finals period was going to be a breeze.

Not so fast. It turned out that the neurostimulator used by the professor, something called a TMS — short for transcranial magnetic stimulator — was not exactly safe to try out at home. It was mostly used in clinical settings, it involved a machine the size of a mini fridge, and there was a very real risk of inducing seizures. Cranial electrical stimulation (CES) involves putting electrodes on your head. The electrodes are connected to a box that controls the electrical current that gets pulsed through the electrodes. The box sets the current to specific amplitudes, frequencies, and waveforms. The problem is that the FDA lists the CES as a class III medical device, which means that a licensed clinician has to administer it. Not a good option for a neurohacker experimenting at home.[1] I looked for other options.

The second dead end was transcranial vagal nerve stimulation (tVNS). It's a new, noninvasive version of the vagal nerve stimulator, a powerful tool that used to require surgical implantation. The vagus nerve plays a key role in your fight-or-flight response and its opposite, the rest-and-digest response. Because of this, researchers believe it could be used to increase attention and alertness or, depending on how you toggle it, relax someone who is struggling with anxiety. Because of its connection to the vagus nerve, there's even a possibility of modulating hormones. No surgery or huge boxes to lug around with you either; you just clip a little plastic dongle onto your ear. The catch? At the time of writing, I found fewer than 20 research studies relating tVNS and cognition.[2] Some new companies were offering consumer tVNS products, and while they sounded feasible, none of them had subjected themselves to rigorous scientific testing by third parties. To me, it's not worth shelling out $200 or more for an untested device. I look forward to testing out tVNS myself once there is more rigorous testing of it.

The third neuromodulation method I considered was photobiomodulation. This one was fascinating, too. The ancient Egyptians used to rely on something like it during mummification. They pulled the brain out through the nose when preparing mummies, and that's because there's actually a direct path through the nose to the brain.[3] Photobiomodulation doesn't pull your brain out your nose (thankfully), but it does involve shining lights of different wavelengths through your nose and up into your brain to modify its functioning. Painless, no surgery, basically just a nose ring. There have been some exciting pilot studies, but it had the same problem as tVNS: not enough research on safety and efficacy yet for me. Also, the at-home units are still hundreds of dollars.[4] Still, I'll watch this space for a few years, and if the research matures, I bet that something interesting will be available soon commercially.

I felt like Goldilocks trying out beds; nothing was quite right. As it turned out, my answer would come through a chance conversation on the way to a British pub, a New Orleans conference, and the unexpected kindness of a group of Stanford grad students in the San Francisco Bay Area.

A Shocking Science

Hoping for a lead on neurostimulation, I'd flown to an educational neuroscience conference at University College London. At the end of the second day, a group of us went to a nearby pub. A Canadian professor overheard me asking a graduate student, "Do you know any researchers using brain stimulation to accelerate learning?" The professor's eyes sharpened. He cut into the conversation, "That's an interesting question. Have you tried Roi Cohen Kadosh at Oxford?" A flurry of emails later, and I found myself walking into a psychology building at Oxford. Within the grandeur of this Gothic cathedral of a university, I would find my next breadcrumb.

"What types of neurostimulation have you looked at so far?" Roi Cohen Kadosh asked. I told him all of the options I'd been through. Did he know of anything safer, smaller, and that even healthy people could use, maybe even students in schools? Something like the device from "Savant for a Day"? He considered for a moment. "Look into tDCS."

Short for transcranial direct-current stimulation, tDCS involves a device that runs positive or negative current through electrodes placed on a specific part of the scalp. The positive electrode—known as the anode—is thought to increase the likelihood that neurons beneath it will fire. The negative electrode—known as the cathode—is thought to decrease the likelihood that neurons beneath it will fire. Where the electrodes are placed affects how current flows and what parts of the brain they affect.[5]

When Cohen Kadosh described the research he was planning—a stimulation experiment where he hoped to increase mathematical ability in people with dyscalculia, the numerical equivalent of dyslexia—it sounded like I had found exactly what I was looking for.

The picture I found on the internet later that night didn't exactly inspire confidence. It looked like a middle school electronics project. There was a 9-volt battery, an electrical circuit the size of my palm, and a couple of sponge electrodes hanging off one side. A little plastic toggle hung off the other side of the circuit; this was to control the amount of electrical current from the battery. Little did I know that I'd be building one myself in a few years.

Fast-forward to the Society for Neuroscience conference in New Orleans a few years later, for which more than 40,000 neuroscientists had descended on the city. This was an especially memorable conference for me, because that was where I met Chung-Hay Luk. Working on her PhD in neuroscience at the time, with an undergrad degree in bioengineering, and a passion for electronic fashion, she had constructed entire ball gowns of twinkling LED lights in her spare time. We discovered our shared passion for cognitive enhancement research, and I admitted that I'd been excited about tDCS ever since that conversation with Professor Cohen Kadosh at Oxford. Also, I admitted that I still hadn't gotten ahold of a device. A grin spread across her face, "Maybe we should just build one ourselves. My place won't work, but I know somewhere that would."

I drove down to the Palo Alto co-op where a group of Stanford grad students were living, with periodic visits from a few of that year's fellows (talented college dropouts attempting to start companies thanks to the largesse of billionaire Peter Thiel). They had generously offered to let us stage our tDCS adventure on their kitchen table. Chung-Hay met me at the door; she wasn't a Thiel Fellow herself, but she was friends with them. Rounding out our trio was Bilal Mahmood, a Stanford biology grad who had recently returned from a White House technology fellowship. I had lured him there with the promise that if we built this tDCS thing he could be the first human guinea pig to try it. He even shaved his head for the occasion. Side note: this actually wasn't necessary—the device works just as well with hair. He'd already done it, though, and I didn't have the heart to tell him. Oops!

Thanks to Chung-Hay's expert soldering and speedy fingers, we created our first prototype in an afternoon. First, we had to check that the device was safe. Using a voltmeter, we confirmed that it produced between 0.5 and 2.0 milliamps (mA), a safe level for humans. Our next step was to design an experiment. I gave Bilal a few different options; he was intrigued at the idea of getting faster at typing. He was a hunt-and-peck typist, he admitted.

As we've discussed, before you do any intervention, you have to establish your baseline. Bilal took a few one-minute typing tests and a reaction time test. Then, we had to pick our tDCS protocol. Since we were trying to improve Bilal's typing speed, we had to pick the correct part of his head to place the anode (where positive current flows into the body) and cathode (where current exits the body). The protocols in the existing research ranged from as little as 5 minutes to as much as 40 minutes of stimulation. We chose 15 minutes. As for placement, we put the anode on the top middle part of Bilal's head, over where the motor cortex of his brain would be.[6] We put the cathode over the orbits of his eye socket, over where the supraorbital region of his brain would be. The electrode sponges were so big that one of Bilal's eyes was almost covered. An awesome '80s-style headband helped keep the electrodes in place.

Final step? Explain to Bilal what was about to happen and make sure he was OK with it. "We're going to zap your brains, buddy. You're cool with that, right?"

Just kidding. What we really said was closer to this: "tDCS has been found to be one of the safest of all neuromodulatory approaches. In a 2016 review article, a group of researchers looked at 1,000 subjects who underwent 33,200 repeated sessions of conventional tDCS (less than 40 minutes at less than 4.0 mA and less than 7.2 coulombs—although most researchers recommend no more than 30 minutes at less than 2.0 mA) and found no reports of irreversible injury or 'serious adverse effects.'"[7]

tDCS adds so little electric current and the voltage is so low (you're receiving far less than the full 9 volts from the battery) that there's no comparison to, say, electroconvulsive shock therapy. In fact, it is so little electrical current, the question is less about the danger and more about whether it is doing anything at all. Just 0.5 to 2 mA of current, spread across a large, spongy electrode and then muffled by your hair, skin, skull, and the tissue just below your skull, is probably not even enough current to directly cause a neuron on the surface of your brain to fire. The theory is that it causes them to fire only if they were already on the threshold of firing. It's more of an electrical suggestion to your brain than a command. The benefit of having both an anode and a cathode is that you can potentially cause a behavior to go in one direction or its opposite. That being said, some users have gotten light burns on their skin or complained of fatigue or headaches.

Shortly after we turned on the stimulation, Bilal made an odd expression.

"Hey guys," Bilal said. "Is it normal for me to be seeing a white light?"

Chung-Hay and I froze. In almost identically strained voices: "What?!" Once we'd gotten the electrodes off him, we demanded details. Bilal assured us that he wasn't in pain. The white flash had gone away as soon as we removed the electrodes. "None of my life flashed before my eyes." He assured us. Panicky laughter.

Chung-Hay recovered first. "OK, I think we just placed the electrodes too close to your eye. It was probably a phosphene." Phosphenes are flashes of light that appear to the person being electrically stimulated. They occur with other types of neuromodulation, too, including TMS. For a while, it was thought that they only occurred when you placed the electrodes near particular parts of the brain, such as the occipital region, which processes visual information, but in 2012, researchers at Rutgers found that they happened more often when electrodes were placed close to the eye. Apparently, the retina in your eye accidentally gets stimulated by the tDCS when the current is too close.[8]

The next few times we ran Bilal through the experiment (yes, he came back even after the white light episode, and with slightly longer hair), we got moderately positive results! His typing speed was faster after the stimulation than it had been before, and his reaction time was better, too.

The Science Behind tDCS

tDCS is still new, but we understand a few things about how it's working in the brain. Enhancement seems to work better when you simultaneously use tDCS and perform the target task you're hoping to enhance performance on. So, any experiments that don't involve simultaneous stimulation and training may be likely to fail by design, which may explain some of the failures we see in the literature. The necessity of stimulation with training seems to be true in other animals, too. For instance, when researchers electrically stimulated a mouse's cortex by itself, using amounts proportional to the amounts used in humans, there was little measurable effect. It was not until another region was simultaneously stimulated that there was a discernible change in the brain.[9]

A group of researchers out of Albuquerque, New Mexico, looked at biological changes in human participants during and after tDCS sessions. They looked for immediate changes in brain waves and the types of longer-term changes that can show up in the brain's white matter.[10] On the immediate side, tDCS seemed to make neurons more likely to fire than they previously would have been. For instance, when tDCS was applied while independently stimulating an arm nerve, the brain wave associated with stimulating the arm nerve was six times higher than before undergoing tDCS. The effect didn't go away immediately, either. Just under an hour after the tDCS ended, independently stimulating the arm nerve still resulted in two and a half times the original brain wave. The big shock came five days later. Brain imaging revealed significant changes in the white matter in the region where tDCS had occurred. That so small an amount of current had rendered a structural change in the brain was unexpected. In mice, tDCS seems to alter glia and calcium concentrations in the brain, and perhaps that explains some of the white matter changes seen in the humans in this study.[11]

Can tDCS Enhance Executive Function?

A 2013 study conducted on 60 healthy subjects in Australia involved participants being randomly assigned to one of three groups: a sham tDCS group in which they played a computerized cognitive training game, an active tDCS group in which they played a computerized cognitive training game, or a group that did tDCS alone (no computerized cognitive training game). The participants had no idea whether they were receiving the real or fake tDCS; when asked to guess which they had received, they guessed no better than chance. All of the participants' executive function abilities got measured three times: before the intervention in order to establish a baseline, after the 10 training sessions were done, and four weeks after the intervention ended. The active tDCS group exceeded the sham group on untrained measures of attention, processing speed, and working memory at the four-week follow-up.[12]

Multitasking: In a 2016 study using tDCS on 20 healthy US Air Force personnel, researchers discovered improved multitasking. The participants first went through a test originally designed by NASA to assess aircraft crew members' ability to juggle tasks such as system monitoring, tracking, scheduling, resource management, and communication. Then one group was given tDCS and another was given a comparable sham treatment. The participants being stimulated with real tDCS performed 30 percent better than those being stimulated with fake tDCS.[13]

Cognitive control: Cognitive control is a key feature of executive function. It's what allows you to flexibly adapt to changing situations—say, inhibit an automatic response to something when you realize it's not adaptive. It comes up when you're pursuing complex goals in the real world.

There is a test of cognitive control called the stop-go task. You respond as fast as you can when you see a "go" signal (a green light) and stop as soon as you see a "stop" signal (a red light). In a 2013 study at Daegu University in Korea, 40 healthy participants were randomly assigned to receive either a real or a sham tDCS intervention. During stimulation, the real tDCS group out-performed the fake tDCS group on a stop-go task. After the stimulation was over, however, the real tDCS group performed the same on the stop-go task as the fake tDCS group.[14]

Working memory: One study published in 2005, another in 2015, and another in 2012 explored the effects of tDCS on working memory.

In a 2005 study led by researchers at Harvard Medical School, tDCS enhanced working memory after just 10 minutes of stimulation.[15] The researchers found that enhancement occurred when researchers placed the anodal electrode over areas of the head that cover the prefrontal cortex and when they placed the cathode over the area of the head covering the primary motor cortex. If they reversed it, it didn't work. This underscores the specific-ity of these protocols, so pay attention when you're trying it yourself; make sure you use the correct configuration for your desired target.

For the second half of the stimulation time (the last 5 minutes), the par-ticipants did a working memory test. An hour later, they received tDCS for another 10 minutes, again taking a working memory test. One of the sessions involved fake stimulation and the other session was real, but the participants didn't know which was which. The researchers also randomized the order (whether participants received active or sham stimulation first). The working memory test was a 3-back letter task. (See page 76 in case you want to try a version of it.) It wasn't a big study; the participants were 15 healthy college-age students (19 to 22 years old). However, it had a positive outcome, and the fact that it was conducted on healthy adults—rather than on a population with a disease, which is often the case with medical research—makes it a usable protocol if you want to try it on yourself.

In 2015, a study was published from a group of researchers at the

University of Pennsylvania who wanted to know whether combining tDCS with a 3-back working memory task (a hard task) or a 1-back working memory task (a pretty easy task) would produce greater gains. They randomly assigned two dozen healthy participants to get either real tDCS or sham tDCS while they did a 1-back or 3-back working memory task. Before and after the combined working memory task/tDCS sessions, the participants took an executive function test. The fastest and most accurate group was the one that did the difficult 3-back working memory task combined with the real tDCS.[16]

This was potentially surprising—after all, that was the hardest task, so they might have been expected to get too tired to perform well. To confirm this, I looked for other studies with similar results. A 2012 study conducted by researchers at University College London had seen a similar effect: after randomly assigning 22 participants to perform a cognitive task with either anodal tDCS or no tDCS, they found that the group with the tDCS performed best.[17]

Can tDCS Enhance Learning and Memory?

In 2015, I attended the Quantified Self conference in San Francisco. One of the talks jumped out at me. College student JD Leadam got up on a podium and gave an impassioned speech about a harrowing time in college. He had struggled for a long time with anxiety, stress, attention difficulties, and learning from texts (he thought of himself as an auditory learner). An adamant self-tracker, he had tried various supplements but found little benefit from any of them. After one sustained session of tDCS, however, he went from convinced he would fail an upcoming exam to acing it.[18] His further experiments and successes with tDCS inspired him to start his own tDCS company.

After the conference, I was curious about how much Leadam's success in using tDCS for improving learning and memory would generalize. Looking into the literature, I found an array of domains where researchers had discovered positive results for learning and memory.

Learning and recall of words from lists: In 2016, a group of researchers from Italy, the UK, and the US conducted a randomized, double-blind study in which one group of participants received real tDCS and the other received fake tDCS. They asked the participants—a group of 28 healthy older adults—to recall a list of words they had learned two days earlier. Then, they asked them to recall as many of those words as they could a month later. Those who had received tDCS while learning the list performed nearly 34 percent better on average (a full standard deviation better) than participants who received the fake tDCS treatment.[19]

Math anxiety: Remember the Oxford researcher Roi Cohen Kadosh who helped introduce me to tDCS earlier in the chapter? When I interviewed him during the summer of 2010, he mentioned that he was curious about how math ability might be enhanced using tDCS. Its benefits were strong with people who had been diagnosed with dyscalculia. He also found that the numerical abilities of 15 university students improved with tDCS treatment—even 6 months later, their improvements remained strong.[20] A few years later, though, he and his colleagues dove deeper and came out with a counterintuitive result: while people with math anxiety did improve their speed on math problems using tDCS, people *without* math anxiety saw their speed temporarily worsened with tDCS.[21]

Learning and recall of faces and names: In 2015, a group of American researchers randomly assigned two dozen healthy adults under the age of 30 to a real tDCS group or a fake tDCS group as they learned a series of face-name pairs. In both immediate and delayed recall situations (measured 24 hours after the stimulation-and-learning session), the group that underwent tDCS performed significantly better than did the group undergoing sham tDCS. How much better? The real tDCS group recalled 50 percent more names, on average, than did the group undergoing fake tDCS. The real tDCS group also made fewer errors in their recall.[22]

Motor learning: If you've ever wanted to learn a musical instrument faster, develop in a sport faster, learn to type faster, or become skilled at other physical tasks more quickly, tDCS may help you do that. In a systematic review and analysis of a little over a dozen articles, researchers found that tDCS tended to improve average motor learning in healthy adults by more than 26 percent on certain types of simple tasks. In other cases, they were much smaller, but in still other cases (such as a few studies on children) the improvements were massive—over 55 percent improvement.[23] However, I strongly recommend you do not try any of this on a child—those studies were conducted under professional supervision.

Can tDCS Enhance Emotional Self-Regulation?

On the advice and information sharing site Reddit, there's a large group devoted to discussing tDCS.[24] One of the most striking sections is focused on the questions and stories of people trying out tDCS at home for their depression. After reading a few of these posts, I began wondering whether tDCS could be used more generally for helping people manage emotions that felt overwhelming. With a little digging, I came across a remarkably thorough and broad review by a group of researchers at universities in the US and Italy, outlined below.[25]

Managing sadness and anxiety: In research conducted in 2006 and 2009, up to 30 days after receiving tDCS stimulation, participants reported that their sadness remained reduced. In one study, participants were better able to successfully identify positive pictures after receiving tDCS, something they had struggled to do when their minds were filled with negative, depressed thoughts.[26] In 2014, high-anxiety participants enjoyed decreased cortisol (a stress hormone) and experienced less anxiety after receiving tDCS.[27]

Managing anger: In a 2014 study, researchers found that when they used a particular configuration of tDCS, they successfully reduced proactive aggression in men.[28] When using the same tDCS configuration as the 2014 study and combining it with a cognitive reappraisal task (a way of reinterpreting or gaining new perspective on a troubling emotional problem), a 2015 study found that 20 participants who underwent 15 minutes of tDCS felt their negative emotions reduced.[29] A 2012 study used the opposite configuration of the tDCS that had been used in the 2014 and 2015 studies (by putting the anode on the left rather than the right side of the head the way the earlier studies had placed it), and participants who had felt insulted responded with more anger.[30]

Persistence and delaying gratification: In 2018, researchers randomly assigned 79 participants to real or fake tDCS for 20 minutes; they found that participants undergoing real tDCS could up their pain tolerance and persist through discomfort for longer.[31] When it came to delayed rewards, another group in 2018 found that two different types of tDCS increased their 24 participants' preference for larger but later rewards.[32] Finally, while it's not emotional regulation directly, it may be worth mentioning a 2008 study out of Harvard Medical School that showed that tDCS seemed to help decrease food cravings and attention to desserts in more than 50 participants. It even decreased consumption, with some benefits lasting up to 30 days.[33]

Can tDCS Enhance Creativity?

In 2007, renowned neurologist Oliver Sacks wrote a piece in the *New Yorker* about a puzzling case that had begun 13 years earlier.[34] An orthopedic surgeon in an upstate New York town named Tony Cicoria had been making a phone call on a public pay phone. The sky was clear but for a few storm clouds on the horizon. Suddenly, lightning shot out of the sky and struck Cicoria through the telephone. While he survived and, remarkably, seemed to bear no significant ill effects, something changed a few months later. Where he had previously been uninterested in music, he suddenly became entranced by piano music. He wanted to listen to it. He wanted to play it. Soon, he began waking up at 4:00 a.m. to play before work. As soon as work ended, he ran back to the piano bench. Not long after, he began composing his own music. Eventually, he was asked to perform publicly. Cicoria's case is not the only one of its kind. There have been more than a dozen verified cases of "acquired savantism," the term used for suddenly becoming advanced at a skill later in life, often brought on by a traumatic head injury.

A few researchers have wondered whether head traumas and lightning might have accomplished the same thing; shutting down or muffling the activity in brain regions responsible for categorizing, judging, and inhibiting other areas. Perhaps those other areas—if given freer rein—could give rise to more creative expression.

Remember the 2003 *New York Times* magazine story that first sparked my interest in brain stimulation? Australian researchers Allan Snyder and Richard Chi conducted an experiment in 2011 that presented participants with a group of problems that involved using a specific set of rules to solve them. Then, right after, they posed a similar-looking new problem set that required ignoring the rules of the previous problems completely.[35] Most of the participants didn't realize this, though, of course. So one group mostly tried to reuse the old techniques to no avail (as a result, only 20 percent of them successfully solved it). The majority of the other group's participants managed to shift their mindset for the new problems, though. They outperformed the other group by almost three times! If you guessed that the group who successfully shifted their mindset to find the new approach was the one subjected to the real tDCS, you are correct.

Returning to musical creativity for a moment, a group of American researchers wondered whether tDCS would affect musical novices and experts similarly.[36] Interestingly, tDCS helped the novice jazz improvisers but hurt the experts. The judges deemed the novice improvisers who had been subjected to the real tDCS stronger than their peers who had been stimulated with fake tDCS. The judges still found all of the novices' improvisations to be of poorer quality than the experts', but the experts who had undergone real tDCS suffered a drop relative to their peers who had been stimulated with fake tDCS. They repeated the experiments over multiple days and sessions and found that the effect held under multiple conditions.

The researchers hypothesized that this occurred because novices tend to rely on controlled, intentional mental processes when they do a task, whereas experts tend to turn off this type of thinking and instead rely on more automatic mental processes. Imagine thinking about whether your right or left leg is supposed to step next when you are walking down the stairs—you'd probably fall! The researchers stimulated the area that runs controlled, intentional mental processes: the frontal cortex. That stimulation supported the type of mental activity a novice needs but not what an expert needs.

Incorporating tDCS into Your Self-Experiments: Some Advice

If you're interested in trying tDCS in your self-experiments, here are a few things to keep in mind. When picking a device, you can build one as my friends and I did, but I'd generally recommend that you buy one instead. Look for safety and customizability. There are reliable devices for under $200, but if you want to run a blinded self-experiment, you'll need a device that includes a sham setting so you can create control conditions. Unfortunately, the ones with this feature are typically made for professional researchers and often go for over $300.

Realize that, as always, self-experimentation will be key to figuring out *your* personal best approach. Individual differences loom large in tDCS findings, so consider the opposite experiences of expert jazz performers versus novices. In some studies I read, whether the participant was right- or left-handed affected the results, in others it was age- or sex-dependent...the list goes on. In my own experiments, one of my friends was very sensitive to the buzzing, warm sensation of tDCS, and the other felt nothing even after we'd gone up to the highest safe level.

Furthermore, what's going on in your mind when you try tDCS matters for the outcome. Jared Horvath, an Australian researcher, authored one of the most cited takedown pieces in the tDCS literature. Years before, when he was working at a Harvard Medical School lab, he and I had discussed tDCS over lunch at the Boston Brewing Company. At the time, he was incredibly bullish about tDCS. However, he was also willing to question even his own enthusiasms. He and his coauthors focused on the effects of a single session of tDCS on healthy adults' cognitive abilities, including executive function and memory. Although for many outcomes there had been only a couple of studies, Horvath got a chance to make an important point: the brain's pre-existing state before treatment affects tDCS's effectiveness.[37] A later paper from a University of Chicago team found that time of day affected whether tDCS made a difference, leading to the hypothesis that "tDCS effects are easier to detect during nonoptimal cognitive processing times (e.g., mornings for younger adults)."[38] We've all experienced that sometimes our brains are ready to learn and sometimes they're not. If the University of Chicago researchers are right, our brains may benefit from the small amount of current offered by tDCS when we're at a relatively low point. This implies that, at least for us night owls forced to operate in a morning-obsessed culture, a self-experiment using tDCS in the morning might be worth considering.

SOME CAUTIONARY WORDS

tDCS is a little more problematic than some of the other inventions since when it comes down to it, you're putting electrical current into your head. Yes, it's a small amount, but any risk needs to be outweighed by significant benefits. This one I've personally used on myself only a few times, not repeatedly like some of the others. I want you to be very careful if you decide to experiment with tDCS. Please watch for side effects. If in doubt, err on the side of extreme caution. Before you start throwing away your coffee and switching it out for neurostimulation, it's worth hearing some additional cautionary advice from Roi Cohen Kadosh, the Oxford professor who originally told me about tDCS.

Concerned about possible "cognitive costs," in 2013, Cohen Kadosh and his colleagues conducted a follow-up study where they used tDCS to improve people's learning rate in one particular type of mathematical ability. When they did that, participants' ability plummeted in a different type of mathematical ability.[39] By placing the tDCS electrodes in a different configuration, they were able to reverse the fall, and the second ability rose back up—but the first ability fell again. In other words, it was a trade-off between the two abilities: they could improve one, but only at the expense of the other. So in all of these tDCS studies in which cognitive gains were made, what silent cognitive cost was being paid without either the participant or the researcher even knowing? If you're trying tDCS, take notes, watch for side effects, and, every few experiments, stop and do a global assessment of your mental performance to make sure that while you were working on one ability, you haven't silently suppressed some other ability accidentally. You want to avoid unknowing trade-offs, or at least address them as quickly as possible.[40]

The second area to keep in mind goes for everything in this book but especially so when it comes to injecting current into your brain. In "An Open Letter Concerning Do-It-Yourself Users of Transcranial Direct-Current Stimulation," University of Pennsylvania and Harvard Medical School affiliated researchers said: "Consider that the level of acceptable risk is different for healthy subjects, who in general are functioning quite well and thus have less to gain—and more to lose."[41] I agree with that sentiment, which is why I have used it so infrequently myself.

Who shouldn't use tDCS? It's worth knowing that many tDCS research labs exclude people with specific characteristics from their experiments.[42] Exclusion criteria range broadly. Labs often exclude people who are pregnant, and anyone with low tolerance for skin irritations or a tendency toward eczema on the scalp. They also exclude people taking psychoactive medications and anyone with depression, bipolar disorder, or any psychotic diagnoses. Finally, they tend to exclude anyone who is prone to seizures, anyone who has non-removable metal in their heads, teeth, or ears (including plates, implants, etc.), anyone with a cochlear implant, and anyone who has been hospitalized for head trauma in the past five years (e.g., for a concussion). The reason is that less is known about tDCS's effects on people who fall into these categories. If you don't have any of these exclusions, it will be easier to predict how tDCS will affect you. That being said, you should still consult a doctor before trying tDCS.

Finally, I'll leave you with a few more general thoughts. As exciting as some of these interventions are—and tDCS is certainly an exciting one—none of these interventions will provide effortless, instant knowledge. They are simply catalysts. You provide the reactants that will create the chemical change that upgrades your brain and various aspects of your life. For instance, if you previously needed to study for two hours and struggled with your focus the whole time, maybe you'll need to study for only three-quarters the time now and you'll find your concentration improved. But you'll still need to study! Remember that research shows the greatest effects of tDCS came when it was coupled with cognitive training—and not just any cognitive training; specifically, the most challenging cognitive training brought the greatest gains. Until the neural implants and full human-computer interfaces arrive, learning will still require work and effort. My hope is that using tools like this doesn't decrease the time you spend learning and stretching yourself; rather, I hope it inspires you to squeeze more learning and growth out of any time available.

TAKEAWAYS

1. Neurostimulation comes in many varieties. One of the safest, cheapest, best-studied, and most accessible varieties is tDCS (transcranial direct-current stimulation). It involves sending a very small electrical current through the skull into the brain.
2. Certain tDCS protocols have helped with executive function, learning and memory, emotional regulation, and creativity.
3. tDCS tends not to have side effects, but when they occur, they include phosphenes (seeing a white light), mild skin irritation, or possibly burns, fatigue, and headaches.

Chapter 19

A Pill a Day

"I learned why they're called wonder drugs—you
wonder what they'll do to you."
—Harlan Miller[1]

We're about to explore one of the most talked-about but also one of the riskiest types of intervention in this whole book. It's one of the only types of interventions we'll discuss that is invasive—it actually goes into your body to do its work. Some people call them smart drugs; researchers and hackers call them nootropics (often pronounced "no-trope-icks"), a term that is said to have been coined by Romanian psychologist and chemist Corneliu Giurgea.[2] Nootropics consist of both prescription and nonprescription pills (supplements, minerals, and vitamins) that can enhance cognitive functions. We'll primarily focus on nonprescription pills.

In this chapter, we'll cover the pills I recommend and also a list of pills I don't. I mention the latter since some of these currently inadvisable pills provide a hint of what we may be able to try in the future—but only if significant amounts of research and development go into modifying them for general use. I also think it's worth knowing what is off-limits—and why. As you advance in your neurohacking journey, you'll inevitably come across people who swear by the (illegal and frequently dangerous) use of prescription drugs. Understanding why they should not be doing this will help you control your own curiosity and keep yourself safe from some potentially horrific side effects. We'll discuss dosages, but always remember that individual reactions can vary widely; even if you use the lowest dose, be on the alert for side effects. Before we go into nootropics I recommend, I want to scare you a little. In a good way.

Nonprescription Drugs, Vitamins, and Supplements Aren't 100% Safe

The FDA has started cracking down on nootropics manufacturers for making medical claims without sufficient research backing.[3] This is not a surprise, considering the industry norms. Regulation and testing of nonprescription drugs, vitamins, and supplements is nowhere near as rigorous as it is for pharmaceutical (prescription) drugs, especially in the US. That means that what you see on the label may or may not be what you get in the box. Also, it could be contaminated with bizarre substances—even poisons like arsenic. Think I sound like a crazy conspiracy theorist? I was skeptical about this stuff, too, until I started researching vitamins.[4] What I found horrified me. Multiple reports indicated that vitamins and supplements often contain arsenic, lead, and other contaminants.[5] One paper explained that the FDA "is limited to removing products proven unsafe, rather than prospectively assessing them for quality manufacturing . . . there is very little FDA oversight until reports of patient harm occur."[6]

The supplement and vitamin industries as a whole—not just nootropics producers—have started to come under fire for the reliability and purity of their products. I used to think that the most likely risk you faced with a "brain-boosting" vitamin or supplement was that it would do nothing. I figured that many would never actually get used by the brain, because they wouldn't be able to cross the blood-brain barrier. Furthermore, the worst case was that if you accidentally took too large a dose, you would just pee it out, right? Scarily enough, in many cases, these assumptions are wrong: while very few compounds can get past the blood-brain barrier, some compounds do stay in your system and accumulate.

In another weird twist, third-party watchdog organizations such as the Clean Label Project, Labdoor, and ConsumerLab have tested consumer vitamins and supplements in their own labs and discovered that, in addition to issues of purity, many of these products do not even contain the amounts of active ingredients they promise on their labels—they contain significantly more or less. Which, as you can imagine, can make dosing quite tricky.

Because the supplements industry is largely unregulated, some companies make scientifically unsubstantiated claims that they can cure serious illnesses like Alzheimer's disease. In 2019, the US Food and Drug Administration (FDA) and the Federal Trade Commission (FTC) warned multiple manufacturers to stop making false claims and warned consumers to be extra careful when considering supplements.[7] A ploy that particularly annoys me is when I read how a supposedly "science-backed" nootropic company has a list of research studies—but they are composed almost entirely of research on nonhuman animals, like mice and rats. Why does that matter? While biomedical research typically starts with tests on nonhuman animals, it must prove efficacy on humans to be compelling. Frequently, an intervention works well on a mouse's brain but, for a variety of reasons, only a low percentage work when translated to a human brain.[8] This failure rate is part of why major pharmaceutical companies have backed away from developing new drugs for the brain.[9]

Another unscrupulous thing that some supplement companies do is claim that they have research backing their nootropic formulas. Often, it's only research on one specific *part* of their compound, not the product as a *whole*. We know that the amount used and the interactions between chemicals used matters a lot. Imagine you read a study that claimed that eating cherry cake would make you live longer. You'd probably be surprised, because cake is not generally high on "healthy foods" lists. You ask how the researchers came to their conclusion. They answered that a study once showed that rats that ate cherries lived longer than rats that didn't. But what about the healthiness of the rest of the cake's ingredients?!

There's another reason herbal, supplement, and vitamin products can end up having low quality or no research backing them: a lack of money. Pharma companies can afford to run expensive clinical trials because patents protect their products from being copied by other firms for at least a few years after they hit the market. Herbs, supplements, and vitamins tend not to be able to get patents. Companies can't afford to put in the money to test something if other companies who don't spend money on those tests can just come in and copy their products.

Are Prescription Pills Safer?

At least in the US, the quality-control mechanisms for prescription drugs are generally stronger than those for nonprescription drugs. That being said, there are no prescription pills designed to make healthy people smarter. That's just not how the pharmaceutical system works; medicines are for people with diagnosable conditions. So, any prescription pill you take may work perfectly for the condition it was designed to treat, but if you don't have that condition, it could actually hurt you.

While "study drugs"—drugs used off-label by people who don't have the condition they're for (for example, people without ADHD using Adderall as a "study drug" to try to stay awake to cram for exams)—are quite popular in some circles, there are some very real downsides to them. For one, this behavior is illegal. The US Justice Department, for instance, considers taking prescription drugs without a prescription to be illegal;[10] this includes taking medicine prescribed to a family member or a friend. There are also some very good health reasons not to do it. Secondhand medication could be old (and some medicines actually become unpredictable as they age) or, depending on the source, could be contaminated with something else entirely.[11] Additionally, the potential for abuse and misuse is high.[12] If you do misuse or overdose on stimulants, the side effects can be brutal; they include hallucination, panic, tremors, stomach issues, and heart problems, to name a few.[13] One fellow approached me at a conference and told me he'd felt "dulled out" ever since he started experimenting with off-label use of stimulant medicines. I believed him; his eyes looked hollow and his voice was flat and expressionless.

Despite some people's beliefs that stimulant medications could improve anyone's mental performance, study after study has found this idea to be, if not fully incorrect, at least incomplete. While people without diagnosed ADHD have seen benefits from amphetamine usage (ranging from memory to mathematical ability), the benefits seem to be much smaller than for people with ADHD. In other cases where the benefits were larger, the researchers noted that their sample may have been unreliable, as they suspected that some of the participants who responded best were hyperactive and simply hadn't been formally diagnosed with ADHD.[14] Yet, many people swear by stimulant "study drugs"; one reason may be that, for those without ADHD, these drugs increase their energy and/or motivation to do things.[15] The drugs make them *feel* quicker and more able to focus. The difference is in the *perception* of their own abilities — not a change in their *actual* abilities.[16]

It's an entirely different story for people who actually have ADHD. The ADHD brain often lacks sufficient norepinephrine, a key neurotransmitter, or its chemical constituents (dopa or dopamine); stimulant drugs can correct this deficiency.[17] A college friend of mine with ADHD had her IQ tested when she was on her medication and off it; she scored more than 20 points higher when she took it. This came as no surprise to the test administrator, who noticed that when she took the test without her meds, she kept pausing to daydream; this led to her not answering some of the questions before the time elapsed. When she was on her meds, however, she was able to focus throughout the entire test-taking session and answer all of the questions. Similarly, when researchers tested the IQs of children with ADHD before they were treated with stimulants — and again a year later — they found that their testable IQs had risen significantly with treatment.[18]

But, again, that was for people with genuine ADHD. Not only do ADHD drugs not seem to provide much benefit to people who don't actually have it, the potential for addiction and other side effects is quite real.[19] So, bottom line: I don't recommend them for people without ADHD. As a neurohacker, you're putting in time and effort to build up your brain to be smarter, better, and faster. Be careful of the risks that you take with that precious organ sitting there, trusting you, waiting patiently between your ears.

And now that my PSA is over, let's dive in!

Who shouldn't try this? If you're pregnant, have a serious medical condition, are taking medications, or are under the age of 25, consider skipping the rest of the interventions in this chapter — or, at the very least, consult with a doctor. The brain doesn't finish a lot of critical development until your mid-20s,[20] and we still have a very preliminary understanding of the neurochemical effects on brain development of a lot of "smart drugs," so you may end up giving your brain its best chance to develop fully if you wait to experiment with any of this stuff until then. If it makes you feel any better, my own brain didn't seem to get its act together until my mid-20s, so, if you feel that you absolutely must get the extra edge, I'd recommend you try the interventions in Part III first and also talk to your doctor. If you're just curious about the rest of this chapter, feel free to read on, but please remember: your young brain is precious—and potentially a lot more sensitive and vulnerable than it will be if you wait a few years.

Herbs for Cognitive Enhancement

Can herbs help your neurohacking journey? Herbal products such as CBD oil, made from herbs in the *Cannabis* genus, have gained a lot of attention in recent years. But humans have been using herbs of all kinds to augment mood and cognition for hundreds, if not thousands of years.[21] Let's take a closer look at herbs: what they are, what they claim to do, and any issues with taking them.

THE RISKS UP FRONT

Always check for possible interactions and side effects and get medical advice (as mentioned before, if you are pregnant, under 25, have a serious medical condition, or are taking medications, consider skipping this set of interventions). A friend of mine had to be hospitalized because of an unexpected interaction with an herb. She had been using a nonprescription supplement (CBD oil) to help her with chronic, low-level anxiety. One day, she was feeling a cold coming on, so she took a regular over-the-counter cold medicine. The next day, she ended up in the hospital from an unexpected (and rare) interaction; months later, she was still having memory issues and brain fog. This is not to demonize CBD—my friend has gone back to using it, just never with cold medicine—but it's worth recognizing that a substance can be harmless in one situation and harmful when combined with something else.[22] Herbs can have potentially dangerous interactions with alcohol and caffeine, too.

Most of the nootropic herbs I'm about to discuss come from long cultural and medicinal traditions, mostly from Ayurvedic and Chinese medicine, and a few from the South Pacific and the Amazon. As well understood and effective as they may be in their countries of origin, these herbs are not fully understood by Western scientists. And even if they work well in India, China, Bali, or Ecuador, we can't be sure that we can replicate the same effects in other countries for a variety of reasons: the herb-derived ingredients might not be manufactured in the same way they were originally—and in the US, these herbs are not regulated, so you cannot always be sure you are actually getting what the label claims you are getting. Additionally, diets and life-styles can differ dramatically from the context in which the herbs were origi-nally used (for example, some herbs work best when digested with other foods that may be common in one culture but rare in another). Finally, the potency level of harvested herbs may have changed due to changes in agricul-tural practices or environmental changes. There are a lot of unknowns.

Nootropic Herbs for Emotional Regulation

While I could not find herbs specifically proven to help with emotional regu-lation, the following herbs may help with aspects of it. We'll discuss evidence that they help to improve mood, decrease stress, and reduce anxiety. When you see the term "subjective well-being," that's just researcher-speak for what most of us think of us as "feeling good, happy, and cheerful."

Rhodiola rosea: This herb was incorporated into Chinese medicine as "hong jing tian," but it's also known as "Arctic root," and it seems to have originated at high altitudes in Europe and then spread across Asia. It grows in some parts of North America, too.

Rhodiola rosea is well demonstrated to energize people who are exhausted. The research I've seen that is most relevant to our purposes is *Rhodiola rosea*'s effect on fatigue. Specifically, it seems to undo the negative effects of fatigue on your thinking abilities.[23] *Rhodiola rosea*'s efficacy doesn't seem to change much if you take it over time, but more research is definitely necessary to know about long-term effects.

If you've got a very stressful job or if you're a chronic procrastinator (i.e., those of us who create our own stressful situations), *Rhodiola rosea* may appeal to you. Imagine a situation where you're not looking to stay awake for just one night but rather if you're looking for high performance over a long period of stress and little sleep. That being said, I didn't see much evidence in the research literature of *Rhodiola rosea* aiding cognition under non-fatigued circumstances. So, I wouldn't assume it would lift you higher than your usual functioning if you weren't already tired.

In *Rhodiola rosea*'s profile on the NIH's National Center for Complementary and Integrative Health site, side effects mentioned included dizziness, dry mouth, or excess saliva production in some research studies.[24] I also saw a few reviews online, where users self-reported that they had experienced an anxiety increase after trying *Rhodiola rosea*. But others said their anxiety decreased.[25] It's hard to know whether the side effects that people report are due to *Rhodiola rosea* or something else they're doing in their lives, because very few people run proper experiments on themselves (unlike you, dear reader, who know to use good neurohacking techniques!). But, there's a much more worrying potential cause: one scholarly report indicated that many commercially available *Rhodiola rosea* products didn't actually contain the amount stated on the label, and they contained varying amounts of key extracts and thus could have unreliable levels of efficacy.[26]

Recommendation: Consult a medical professional.

Ashwagandha: Ashwagandha is one of the better-known Ayurvedic herbs. Like many traditional herbs, it is used for many purposes. We will focus on its ability to fight stress and anxiety.

For stressed-out participants given 300 mg of ashwagandha daily, multiple effects have been found. In a 2012 study, after 60 days of the dosage, the participants reported in a general health questionnaire that their social dysfunction had reduced by 68.1 percent, compared to a 3.7 percent reported decrease in the placebo group.[27] In the same study, participants getting ashwagandha reported that their stress had reduced by 44 percent, while the placebo group felt only a 5.5 percent reduction. It's worth noting that the study was double-blinded, meaning that at the moment participants received their pills, neither the participants nor the researchers knew who was getting placebo pills and who was getting true ashwagandha pills.

Another study of stressed and anxious participants who took 300 mg of ashwagandha, this time twice daily with 1.5 percent withanolide, found that when participants took ashwagandha alongside counseling and learning breathing techniques, their symptoms improved by 56.5 percent. The placebo group also received counseling and breathing technique instructions; they improved by 30.5 percent.[28]

Different doses of different ashwagandha extracts had positive impacts as well. A healthy but stressed group of women and men took 250 mg twice daily of a different type of ashwagandha extract, the ethanolic root extract, for eight weeks. The 39 ashwagandha takers reported less stress than did the 19 in the placebo group, and the ashwagandha takers also had improved sleep quality.[29]

Another dosage and type of extract showed promise in reducing the symptoms of stressed people, too. With two divided daily doses for six weeks of 125 to 250 mg of ashwagandha (the extract had 11.90 percent withanolide glycosides, 1.05 percent withaferin A, and 40.25 percent oligosaccharides with 3.44 percent polysaccharides), participants reported significant reductions in anxiety and related symptoms such as lack of sleep, forgetfulness, and other issues.[30]

As is typical for traditional herbs, I couldn't find research studies that followed consumers over the long term to see how well they did and if they ran into safety concerns.[31] We need more research.

At the dosages mentioned above, and if taken short-term (weeks to a few months), no major side effects were found. There are lethal dosages if you take too much, though, so be careful. Even at low dosages, there can be minor side effects including lightheadedness, headaches, sleepiness, and stomach upset. In rare instances, ashwagandha takers have reported elevated liver enzymes and liver injury, sometimes an allergic reaction, or a rapid heartbeat. Ashwagandha can lower blood pressure and blood sugar, and it can increase thyroid hormone levels, so if you have preexisting concerns on any of those scores, consult your doctor before experimenting with it.[32] To help avert stomach upset, at least, taking ashwagandha with a meal may help.

Recommendation: Consult a medical professional. For dosage recommendations, Examine.com is a reliable, research-based online resource. Their page on Ashwagandha is: examine.com/supplements/ashwagandha/#how-to-take.

Kava: Ever been to a kava bar? They are a growing trend in the US. My trip to a kava bar in San Francisco went like this: My friends and I ordered a shell of kava each. The bartender waxed euphoric about the liquid's health-giving benefits and its ancient role in religious and ceremonial traditions in Fiji as he poured a muddy brown liquid into three half coconut shells. The taste was not surprising, given its appearance: it looked like muddy water and it kind of tasted like it, too. Things got interesting just a moment later, though. My tongue and cheeks started to go just a little bit numb. We asked the bartender about the numbness, and he assured us this was a normal effect of the drink. Very soon a relaxation spread over me that I hadn't felt in a long, long time.

It reminded me of that relaxed but alert happiness unique to the first week or so of summer vacation as a kid. My friends commented that it was a bit like drinking alcohol (obviously with a different taste), but that the mindset it put you into was decidedly different. My mind felt surprisingly sharp in a way that even caffeine hasn't done for me. I felt more curious, open-minded, talkative, playful. I asked the bartender if people ever got belligerent after drinking kava the way they do with alcohol. He laughed, as if the idea of hurting another human being would be the last thing a kava drinker would ever do.

Later, as we got up from our booth, I was surprised to notice something had changed in me physically. Despite my continued mental sharpness, the rest of my body didn't seem quite as responsive. "I guess we shouldn't try to drive or operate heavy machinery after kava," I joked. I definitely didn't feel drunk—no issue with walking or coordination—but I did sense that my physical reaction time might be slowed down.

I dove into research as soon as I got home. Semi-scientific blogs pointed out that kava was associated with an "increase in subjective well-being" and a "decrease in anxiety." Yup, sounded about right.

In a 2012 study, researchers randomly assigned participants to take one dose of 30 mg oxazepam, which is a prescription anti-anxiety drug, or one dose of a placebo drug, or one dose of a pill version of kava (180 mg of kava-lactones). Participants didn't know which type of pill they were being given each time, but they took cognitive tests before and after taking each pill. Each session was separated by a week, so they got to try each of the three pills. Kava's performance turned out to be squarely in the middle of the pack. While people's anxiety levels tended to rise after they took the placebo pill, their anxiety remained at the same level after taking the kava pill. After the oxazepam, people's anxiety levels decreased on average, but alertness suffered. Not so after the kava, as I'd felt at the bar.[33]

In a 1997 study out of Jena University in Germany, a double-blind, randomized controlled trial conducted over 25 weeks (roughly 6 months) randomly assigned 100 participants diagnosed with anxiety to kava or a placebo. The kava group began to experience reductions in their anxiety in a significant way starting at week 8 (roughly two months in).[34] Researchers at the University of Melbourne reviewed studies of randomized controlled trials with multiple weeks of kava use for people diagnosed with anxiety (mostly generalized anxiety disorder); there were significant results, with nearly a 36 percent reduction in symptoms in four out of the six studies reviewed.[35]

However, there have been some serious concerns about liver damage and toxicity when kava is taken over the long term or when taking a single large dose. While most of the concerns haven't been clearly linked to kava use (in multiple cases, people were taking other medications or substances that may have caused the problems), some sources still recommend that people wanting to use kava consistently should have frequent tests of liver function. In addition, some people reported skin rashes, but again, that was not proven to be directly caused by the kava. Also, while kava is unregulated in the US and in many other places, it still requires a prescription and is subject to tight regulations in some parts of Europe.[36]

Recommendation: Consult a medical professional. A World Health Organization report warned that it's very easy to consume more of kava's active compounds when drinking kava than when taking the pill form.[37] Maybe on the weekends, you could find a kava bar as a more intellectually engaging alternative to the standard bar. Note that bars vary in the amount of kava they serve in each shell. For more dosage recommendations, especially for experimenting with it in pill form, the kava page on Examine.com is at: examine.com/supplements/kava/#how-to-take. To stay on the safe side, I would recommend bringing a designated driver and perhaps limiting yourself to just one shell per session. Have your liver function tested routinely if you decide to use kava more than this.

Nootropic Herbs for Executive Function

That coffee you drink, that tea you sip, that chocolate you savor, that soda or energy drink you chug all contain one of the oldest nootropics known to humans: caffeine. Caffeine has been used by cultures all over the world for thousands of years.[38] While caffeine by itself has pros and cons as a nootropic, when combined with L-theanine, an amino acid found in tea, some of the weaknesses fall away.

Most people consume caffeine for its energizing power, assuming it can do nothing more. I love the coffee mug that says, "Coffee: do stupid things faster and with more energy!" Certainly, it does up mental alertness and speed, but those are not the same as intelligence. As any college student pulling an all-nighter can attest—and as neuroscientists have confirmed in the lab—caffeine is great for keeping you awake when you're sleep-deprived, too.[39]

Still, there's truth to the criticism implicit in caffeine helping you to "do stupid things faster and with more energy." In dozens of experiments, caffeine at levels ranging from 50 mg of caffeine (roughly half a cup of coffee) to 450 mg of caffeine (roughly four cups), researchers have shown that caffeinated participants improve on the number of hits they can make on simple tasks, and they can improve their reaction time—especially on tasks requiring close vigilance. However, caffeine often impairs participants' performance on more complex cognitive tasks—for instance, those involving memory.[40] Interestingly, one group seemed to enjoy a genuine boost to their raw cognition: extroverts. In a 2010 study out of Goldsmiths, University of London, caffeinated extroverts—but not caffeinated introverts—enjoyed a boost in their working memory.[41]

Our friend L-theanine comes in as the wise sidekick who can harness caffeine's manic energy into something smoother and more directable. Ever noticed that drinking a cup of green tea tends to be energizing but doesn't give you the jitters that coffee does? That's the L-theanine in action.

Users of combined caffeine and L-theanine, like users of caffeine, reported lower fatigue, higher alertness, less of a tendency to get headaches, better reaction times, and even improved speed at word recognition tasks.[42] Additionally, from the research so far, it looks like combining L-theanine with caffeine improves people's ability to control their attention (switch it from one task to another as needed), reduced their susceptibility to distractions, and improved their focus overall.[43]

Side effects of caffeine alone—which can be mitigated by adding L-theanine—include jitteriness or anxiety, "caffeine crash" (a lot of energy at first and then a sudden loss of energy as the body metabolizes it), headaches, nausea, increased blood pressure, restlessness, flushed face, insomnia, cardiac arrhythmia (or irregular heartbeat), muscle twitching, gastrointestinal problems, and a rambling flow of thought and speech.[44]

Recommendation: Consult a medical professional. If you want the cognitive enhancement effects, studies that attempted to study tea levels found very little benefit unless the caffeine and L-theanine levels were both higher than you would get from a regular cup of tea. In this case, you're better off combining coffee with an L-theanine supplement.

Many of the positive cognitive effects disappear the more you use caffeine. To avoid developing a tolerance, you can cycle caffeine. If you're a heavy caffeine consumer (more than 4 cups of coffee per day), to get back the cognitive benefits of "naïve" caffeine use, you may need to quit coffee for up to a month.[45] Although cognitive benefits wane with sustained use of caffeine, its ability to wake you up remains. Or, maybe it's just reducing your withdrawal symptoms. After all, it's possible that you feel good after that daily cup of coffee not because the coffee is making you feel better but because you no longer feel the ill effects of being in caffeine withdrawal.[46]

If you're new to caffeine, you'll probably want to start at 100 mg. That's about as much caffeine as a cup of coffee. A good dosage is 100 to 200 mg L-theanine combined with 100 to 200 mg caffeine (some swear by the ratio being skewed toward more L-theanine, others toward caffeine). The effects seem to disappear in research studies using lower dosages than about 100 mg of L-theanine.[47] Don't overdo the caffeine, because people have actually died from consuming too much.[48] The maximum caffeine to use per day is around 400 mg.[49] If you're pregnant, don't exceed 200 mg per day. Your genetics help determine your responsiveness to caffeine; the CYP1A2 and AHR genes, among others, regulate this, so you can check 23andMe or other consumer genetics sites to see if you have access to information on caffeine-relevant parts of your DNA.[50]

Nootropic Herbs for Learning and Memory

Bacopa monnieri takes patience—up to 12 weeks of taking it before the memory benefits really kick in—but it may pack the biggest cognitive enhancement punch of any of the nootropics. Also known as "brahmi," "herb of grace," "water hyssop" and "Indian pennywort," this plant grows in wet, tropical areas all over the world and has probably been used as a medicinal herb for thousands of years.

When you first start taking *Bacopa monnieri,* its primary effect is apparently in reducing anxiety. It's not until 8 to 12 weeks later that the memory-enhancing effects kick in.[51] (Some people report them as soon as 3 to 4 weeks, though.) *Bacopa monnieri* competes well compared to other herbs and even prescription medicine. In a 2013 review of double-blind, placebo-controlled studies comparing the effects of *Bacopa monnieri,* modafinil (a prescription drug prescribed for people with narcolepsy or shift work in order to stay awake, which has also been used as a nootropic by some), and panax ginseng, *Bacopa* came out very favorably. Although it didn't help participants better remember information they had learned *before* they started taking *Bacopa monnieri,* and it also didn't help them learn new information *faster,* *Bacopa* did help them hold on to the new information for longer—essentially, they forgot less. The result was not subtle, either. In fact, those taking *Bacopa* performed nearly 35 percent better than placebo on cognitive tests; Ginseng and Modafinil came in around 30 percent.[52]

In another 2013 review of over nine studies including 518 subjects, the results came out consistently strong on two other cognitive tests: a measure of multitasking and visual attention (the trail test B) and a type of reaction time test (choice reaction time).[53] While the total number of studies on *Bacopa monnieri* is still low, the quality and persuasiveness of the evidence are growing.

There are contraindications/side effects: some biohackers report nausea, cramping, bloating, and diarrhea,[54] which appear to be mitigated by taking it with some fat. Traditionally, *Bacopa* is served with ghee (clarified butter).

Recommendation: Consult a medical professional. *Bacopa monnieri* is relaxing, and some even say it demotivates or makes them lethargic after taking it, so you may need to pair it with caffeine or just take it at night. Side effects could include nausea, cramping, bloating, and diarrhea. Also, to make sure it's absorbed properly—and doesn't cause too much stomach upset—have it with food. Ideally, you should eat something with some fat in it. Because of how long studies have shown it takes to see changes to memory and learning, you'll want to use the protocol where you compare your baseline results on learning and memory tests before to your results after taking *Bacopa* for 8 to 12 weeks. For specific dosage recommendations, refer to the *Bacopa monnieri* page on Examine.com at examine.com/supplements/bacopa-monnieri/#how-to-take.

Nootropics for Creativity

When it comes to nootropics that enhance creativity, the research is thin. There are a few candidates, but I don't feel confident recommending any of them. If you're of an age that it's legal for you to drink it, alcohol in low amounts may enhance creativity. There was one study in which researchers got their participants slightly sloshed[55] and had them do creativity tests; the intoxicated participants performed slightly better than the participants in the control group, as long as they didn't drink *too* much.[56] Since alcohol helps reduce anxiety, inhibitions, and self-consciousness, it's likely that alcohol reduced the internal "editor" that can hold back creative thinking.

It's also quite possible that the nootropics in the emotional regulation section—*Rhodiola rosea*, ashwagandha, and kava—could help boost creativity by allowing you to manage your moods better, especially since we know that a strongly positive mood (or a negative but calm mood) can improve the odds of being creative. Still, I've not seen any formal studies examining the effects of any of the above herbs on creativity—so, maybe you'll conduct the first on yourself!

Some of My Favorites

There are other beverages that I enjoy for their ability to enhance my focus, boost my mood, and up my energy. They're legal and I've not experienced any negative side effects, but they don't yet have much scientific evidence behind them yet. Given the lack of conventional scientific evidence, I'm hesitant to strongly recommend them. On the other hand, it would be hypocritical not to mention ones that I use frequently myself.

The first is guayusa. This Amazonian tea has about as much caffeine as coffee but with much less jitteriness and way more antioxidants than almost any other tea. It has a mood-lifting effect (at least in part due to its caffeine content), and I find myself more creative on it, too—I find it perfect for writing.[57] It comes from the rainforest, primarily Ecuador, and drinking it may help encourage more rainforest to be preserved and even replanted, which, as celebrity guayusa endorser Leonardo DiCaprio and others have pointed out, would be a wonderful—and much needed—move for the environment.[58]

I've also enjoyed the following for their effect on focus: yerba maté, lion's mane, and chaga root.[59] The first, I typically drink as a tea; the last two I add to coffee. While some of these plants have been tested on other animals, and some of their biochemistry has been analyzed, we have a long way to go in understanding how they generally affect humans' mental performance.

The Also-Rans

These are the nootropics that are worth mentioning but probably not worth trying.

Nicotine: This is one of the oldest and most controversial of nootropics. We absolutely know that smoking causes cancer, and we typically think of nicotine when we think of smoking. While nicotine is a main ingredient in cigarettes, and it can be addictive by itself—apart from all the additives and compounds that it typically gets packaged with in tobacco-based products— the research on nicotine shows it to be more of a nuanced bad guy than an absolute villain. For instance, researchers in the 1980s found that an oddly lower percentage of smokers than nonsmokers developed Parkinson's.[60] In a meta-analysis of more than 40 clinical studies on nicotine and smoking, nicotine, like other stimulants, was found to improve alertness, attention, episodic memory, and working memory.[61] Does that mean we should all go out and buy a pack of cigarettes? Definitely not. However, this research gave me an idea, a dangerous idea: Maybe I could get some of the benefits of nicotine without the downside?

At the drugstore, I felt rather sheepish buying nicotine gum, lozenges, and patches. After all, I've smoked a grand total of two cigarettes in my life, and there were people who were struggling to get rid of the addiction; this felt a bit disrespectful. Still, I decided to go forward with my experiment.

First, I started with a nicotine patch. That was an immediate failure: within a minute, my skin felt like it had caught on fire and was visibly turning red. OK, patches were out. What about the lozenges? The next day, I did a series of cognitive tests, sucked on a nicotine lozenge, then took another series of cognitive tests. No ill effects but no discernible benefits either. Also, the lozenges tasted terrible, and I couldn't see myself using them again. OK, what about the gum? This time, I'd found something I could tolerate.

The taste was a little sharp but not bad. I did my series of cognitive tests, chewed on the nicotine gum, and then took another series of cognitive tests. This time, I saw a significant uptick in multiple measures, especially in verbal fluency (my ability to come up with words starting with a particular letter under a time limit). Additionally, I felt more alert—almost like I'd had a coffee, but a more pervasive alertness that filled my whole body. Maybe I was onto something.

I tested it again the next day, but I skipped breakfast and also drank a coffee right before. This, as it turned out, was a horrible idea. Very soon, I found my vision narrowing, a sense of doom, and incomprehensible fear coming over me, and my hands visibly started shaking. Realizing something was wrong, I called my mother, who is a physician. She asked me what I had been up to (she knows about my neurohacking experiments) and quickly recognized that I was having a panic attack. Just knowing that was what I was experiencing helped calm me down significantly. I went to the kitchen, got some food, did some breathing exercises, avoided caffeine for the rest of the day, and felt better within half an hour. I resolved to be much more careful when combining stimulants, especially on an empty stomach.

I decided to keep nicotine gum around for "emergencies"—times when I really need to focus and I know I'll only be using it temporarily. Practically speaking, though, I end up using it once a year at most. There are other nootropics I've found to be nearly as effective and far less scary that I much prefer. Other friends of mine had similar experiences with nicotine-induced panic attacks—one even ended up with severe withdrawal symptoms from nicotine gum after chewing it every day that lasted for months.

Racetams: Racetams are variously described as drugs, supplements, or simply "agents." They seem to modulate neurotransmitters such as acetylcholine and glutamate, but the exact mechanism by which they modify brain function is not well understood. When you scroll through online nootropics forums or you interview neurohackers, you're bound to come across someone extolling the benefits of racetams, including piracetam, aniracetam, oxiracetam, pramiracetam, and phenylpiracetam. Some individuals swear by them, but the research literature is so mixed as to their efficacy that I have never tried them myself. A few studies showed cognitive boosting effects on people with cognitive differences or impairments (older adults suffering with mild cognitive impairment as well as one study showing benefits to people with dyslexia),[62] but nothing definite for people deemed generally healthy and of average or higher functioning.[63] In fact, the FDA has started cracking down on nootropics manufacturers for making medical claims without sufficient research backing, including those mentioning racetams in their ingredients lists.[64]

Modafinil and armodafinil: These increase how alert you feel and also seem to have executive function-enhancing effects.[65] They are typically prescribed to help people who must stay awake for long periods of time (such as long-distance drivers or pilots) or for people who struggle with constant sleepiness (such as those with narcolepsy). The drug maker, Cephalon, reports an incredible list of side effects: 1 to 10 percent or more users get chest pain or palpitations, headaches, nausea, sweating, changes in their urine (such as bloody urine), back pain, chills, anxiety, and/or insomnia.[66] And those are just the common ones; the less common ones include decreased libido, rashes, amnesia, and suicidal ideation (in 0.1 to 1 percent of users). Modafinil is also reputed to render pretty intense hangover effects when you come off it. So I really wouldn't recommend trying these if you don't have narcolepsy.

A Potential Nootropic to Watch

I would be remiss if I didn't mention psychedelics in a chapter on nootropics. Microdosing psychedelics involves taking 5 to 10 percent of the usual dose of LSD or magic mushrooms;[67] and it has become popular for improving mood and reducing anxiety in some circles.[68] In the fall of 2020, Oregon became the first state in the US to legalize psilocybin, the active ingredient in magic mushrooms, in mental health settings.[69] When it comes to its use in cognitive enhancement, however, the data are less promising. The few studies that do exist have mostly shown mild or no effects.[70] To be fair, this isn't completely damning—after all, conventional research tends to average effects across many people and thus can ignore positive effects experienced in some individuals. Still, there is significant risk associated with using less rigorously tested interventions, especially if something is illegal in the place you are using it. When choosing interventions for your own neurohacking journey, there are plenty of other options that are more broadly legal and have been more thoroughly vetted scientifically.

TAKEAWAYS

1. Buyer beware: nonprescription vitamins, supplements, drinks, and herbs aimed at brain health and enhancement are not regulated in the US (and many other countries). Relatedly, what ends up in any particular bottle you buy will not necessarily be pure. Also, there can be significant variability in how much of a compound you're getting, making dosing tricky and increasing the potential for unknown interactions.

2. Prescription pills are for the illness they were tested on; take them only if prescribed to you. Otherwise, you could very well end up with bizarre or even dangerous side effects. Prescription pills can make you feel more *confident* (as in the case of Adderall), but often they don't actually make you more *competent*.

3. Herbs have uses for emotional regulation, executive function, learning, and memory, but seriously consider skipping these if you're under 25, are pregnant, or have a serious medical condition. Also, consult a medical professional when deciding on dosing.

Chapter 20

Sci Fi to Sci Fact

*"Any sufficiently advanced technology is
indistinguishable from magic."*
—ARTHUR C. CLARKE

> **Time Investment:** 33 minutes
> **Goal:** To understand the power of neurotechnologies
> coming down the pipeline

Before you embark on your own neurohacking adventures, I can't resist leaving you with a few final topics to chew on. Aren't you curious about what kinds of brain-related technologies are coming down the innovation pipeline?

We'll devote the first part of this chapter to exploring technologies that may sound a lot like science fiction. Yet, they are fast becoming—or are already—scientific facts.

We'll cover three categories of upcoming technologies: genetic selection and modification, becoming hybrid humans (part human, part machine), and advances in cognitive data. We'll also look at a few of the larger societal questions that you may want to consider as you define your own personal neurohacking ethics. Just because you *can* try something doesn't mean you should. Neurotechnology could exacerbate the divisions between haves and have-nots in our world. What if the haves get early access to technologies that allow them to not only have more but become more—whether that means smarter, happier, or able to live longer? After that, we'll discuss the neuroscience-related technology advance I'm most excited about.

Smarter Genes

Is intelligence genetic? Can problems you might have with learning go back to childhood, or even earlier? And if so, what can you do about it? One question I had was: Were my childhood reading problems at all genetically based?

It was 2011, and I had decided to do a swab test using the company 23andMe's kit to find out more about my DNA generally. When I saw that there was information on whether I carried genes that predisposed me to dyslexia, I clicked through eagerly. The answer? No. What else could I learn about my mental genetics? Apparently, I possessed a gene that increased my chances of having a higher-than-average episodic memory. This set me to recalling times my memory of a shared event had been sharper than friends' or family members'. Soon, I was nodding, feeling seen. Other results matched my personal experiences, too. I stopped. This was starting to feel like a high-tech version of a horoscope. How accurate was all this?[1] In the case of the cognition-related genes, most of the science is still very preliminary. I still check my 23andMe results every year or so, and I often find that the interpretations of my DNA have changed. For instance, 23andMe once predicted that I was a night owl, but the most recent version predicts that I'd prefer waking up early (nope, got it right the first time!). In many cases, researchers are still trying to figure out which genes help give rise to which behaviors or characteristics because the science is still very much in development.

How much do genes affect cognition and achievement?

The last few decades have exploded the amount of genetic data we have available. We had the billion-plus-dollar Human Genome Project and all sorts of other amazing breakthroughs — and yet, extracting truth from genetic information has been a lot harder than researchers initially expected. We have seen many tests on adults and children trying to get to the heart of this question: Do participants with higher scores on intelligence tests and/or with more advanced degrees or developmental milestone achievements (among kids) tend to share the same genes or combinations of genes?

At this point, the answer seems to be maddeningly inelegant; it looks like the genetic contribution to intelligence is *not* a single-gene job. Rather, it seems that there is a large number of genes, each seeming to contribute to intelligence to varying degrees (as measured by imperfect IQ tests).[2] To date, researchers have sifted through millions of data points, and it appears that many combinations can lead to higher cognitive ability. A recent review that amassed data from "cognitively healthy individuals" gathered between 1995 and 2009 found that over 50 genes seemed to play a role in intelligence, but the estimated effects of each gene ranged tremendously. Whatever role genes play in intelligence, it is likely a coordinated team effort, not the work of just one gene.[3]

To explore the roles of nature and nurture in how much someone achieves, we can look to Swedish psychologist K. Anders Ericsson, the originator of the research behind the "10,000 hours" rule that Malcolm Gladwell popularized in *Outliers*. Ericsson has devoted his professional life to studying the root causes behind the success of world-class performers, whether they're musicians, athletes, or gamers. He takes an environmentally focused stance: "Since we know that practice is the single most important factor in determining a person's ultimate achievement in a given domain, it makes sense that if genes do play a role, their role would play out through shaping how likely a person is to engage in deliberate practice or how effective that practice is likely to be. Seeing it in this way puts genetic differences in a completely different light."[4]

My bet is that he's on to something. If we can find a small set of "stick-to-it-iveness" genes, they'll probably play an outsized role in overall achievement, although it will be interesting to see what aspects of mental performance they factor into. After all, you've probably met someone who scored high on IQ tests and did well in school but then struggled when faced with projects that required longer hours or less structured work. My guess is that some genes will predict to some degree how well you do on standardized tests, and a different set will predict how well you can accomplish long-term endeavors like building a company or raising a family.

CAN WE MAKE SMARTER BABIES?

This question periodically pops up in discussions of emerging neurotechnology. While researchers haven't found a single "smart" gene solely responsible for boosting cognition, they have found the opposite: specific genetic variants that lead to *impaired* cognitive functioning. Globally, it has become increasingly routine to get screened during pregnancy for Down syndrome, neural tube defects, trisomy 18, and Smith-Lemli-Opitz syndrome, all conditions that come with, among other difficulties, a significantly increased risk of intellectual disability.[5] A stark decision arises about continuing with the pregnancy. For that reason, some couples with a family history of specific conditions have sought out reproductive technology like IVF (in vitro fertilization) to help them select eggs and sperm with a lower risk of carrying the genes for incurable conditions.

Does this mean that kids with eerie levels of cuteness, bulletproof good health, and staggering intellect will be filling up your neighborhood playground sometime soon? Probably not. Certainly, reproductive technology *could* be used for certain types of nonmedical enhancements — to select for the baby's eye color, for instance — but very few traits are well enough understood that a simple genetic selection could be done in this way. More broadly, genetically modified babies are not only discouraged, they are actually outlawed in many places.[6] As new technology enters the picture, though, some of the old rules may be reexamined.

In 2018, rogue Chinese scientist He Jiankui used a genetic editing technology called CRISPR-Cas9 in an attempt to confer HIV resistance to twin baby girls. The global response was very negative; Chinese authorities fined He the equivalent of $430,000 and sentenced him to three years in prison.[7] Countries around the world have come down against this type of behavior, some more firmly than others.[8] During the summer of 2019, the US House Appropriations Committee voted to continue the US ban on genetically modifying babies.[9]

Despite the blowback, CRISPR is seen as a revolutionary genetic tool. Back when I worked in molecular biology labs, it was not possible to genetically modify human genes. To explore gene editing, we worked with "genetic model organisms," such as mice. We had to use a much more complicated approach that involved designing a specific protein for each DNA sequence we wanted to change. This was time-consuming, error-prone, and expensive. Using CRISPR-Cas9 (and other CRISPR systems) is easier, faster, and cheaper. These systems allow researchers to hijack the natural defense mechanisms that bacteria use to fight off viruses. Knowing about these systems has allowed us to turn bacteria into little machines that do our bidding; now, we can search for target sequences in DNA and replace current sections with new, desired genetic sequences. Then, the machinery adapted from bacteria produce the proteins we want. This has already started to revolutionize genetic research in many fields, including neuroscience.[10]

Still, when I went to one of his 2016 talks, MIT professor Feng Zhang, one of the leaders of CRISPR-Cas9's development, noted its shortcomings; he pointed out that it can sometimes snip segments of DNA that were not intended.[11] At the time, Zhang was working on a technique to up the specificity for a more precise protocol, enabling us to better understand—and perhaps one day treat—complex diseases with multiple genetic and epigenetic causes. Since cognition clearly has multiple genetic and epigenetic contributors, it's not too much of a leap to imagine that CRISPR-Cas9 will help us decode this too.

If you want to play with CRISPR-Cas9 yourself—by editing the genes of some bacteria, not editing your own genes, mind you—there are home kits available for less than $170 from an online startup called the ODIN.[12] However, the protocol involves keeping *E. coli* bacteria in your fridge. I wasn't thrilled by the idea of my food potentially getting contaminated by *E. coli*, so I elected to hold off on this particular home experiment.

Hybrid Humans: Machine-Aided Cognition

When people envision a future with cyborgs, some may naturally jump to scenarios with them taking over the world or AI overthrowing humans entirely. I'd like to paint a few scenarios in which technology informed by neuroscience and cognitive science could actively reduce and even prevent human suffering. Let's explore a few places where blurring the division between human and machine might actually be worth it.

REDUCING HUMAN ERROR

Drowsy drivers, exhausted surgeons...how many lives have been lost due to human error? A variety of car models now have in-car "drowsiness detection systems" that use eye-tracking systems to make sure that your eyes remain open while you're driving.[13] Gmail used to have a feature called "Goggles" that prevented late-night drunken emails that you later regretted. If it was past a certain hour, it asked you some basic arithmetic questions. If you got too many wrong, it closed you out of email.[14]

What if you had a personalized system that tracked your physical well-being throughout the day and gave preventive recommendations? For example, tiredness often causes irritability and poor judgment. Rather than randomly yelling at your kids and spouse at the end of a long day, you could take preventive action to curb your petulance hours before. A drowsiness detection system could continuously monitor your heart rate, eye movements, respiration, and other biological indicators. Tired people rarely realize how tired they are until they're staring at the result of their own mistake or out-of-character behavior. If an objective monitoring system predicted that you would be dangerously exhausted in a few hours, you might be more willing to take a preventive 10-minute nap or a brisk walk during your lunch break.

In cases in which most humans would struggle to plan, strategize, make decisions, adhere to previously made commitments, or handle extraordinary amounts of new information, technology can help. Max Planck Institute researcher Falk Lieder studied the Clearer Thinking website's "Decision Advisor" tool and found that "participants experienced between 27 percent and 38 percent less regret about their decision after using the tool than when they made a major decision without the tool."[15] The tool asks very specific free-response questions to help participants avoid common cognitive biases, such as forgetting to consider alternative options.

Similarly, in cases in which our natural wiring as humans is likely to lead us astray, having a little cognitive support at just the right time could be life-altering. Our brains have not evolved drastically in the last 20,000 years, when our brains focused on avoiding predators, staking out potential prey, and noticing and gathering nuts, berries, and other treats. Our attentional systems prioritized novel and fast-moving information and issued extra energy and focus whenever we were in those modes. Pairing a 20,000-year-old brain with many of today's tasks is a recipe for error, since much of modern work is repetitive and detail-oriented. Data collection (where the same question is being asked again and again, such as maintaining customer information in retailing) and highway driving are two examples of tasks that can be tedious and repetitive—exactly the sort of job where the human brain is likely to make an unforced error or fall into microsleeps.[16] Total automation of work raises questions of taking away the entire livelihood of a person, but just the right amount makes our work lives more successful and enjoyable.

Self-driving cars, unlike humans, don't get tired, give in to road rage, or take their hands off the wheel to answer a ringing cell phone. Humans and AI together, however, could offer the safest driving experience of all—one in which the inherent weaknesses of our human executive functions are shored up by the strengths of the machine and vice versa.

Everyday helpers

Another small way in which technology could aid with executive function is to help make everyday tasks easier: we may never forget our keys again. In Sweden, thousands of people have elected to implant microchips in the loose skin between their thumb and their palm.[17] These chips, which often employ radio-frequency identification (RFID), allow people to open the doors to their homes, offices, and other locations using nothing more than a wave of the hand. Since Sweden's largest train company allows riders to use their chips instead of tickets,[18] many expect other uses to follow. Already, some have started storing other data in their chips, including medical IDs. The rising popularity of cashless credit systems has led some to think that money may be the next item stored perpetually between thumb and palm.

So far, there seem to be no issues with the chips getting magnetically attracted while going through airport security or when using medical imaging such as MRI and, if implanted properly, there is little risk of infection. Since a would-be thief would have to know you had a microchip in your hand, the risk of theft is probably similar to the risk of having your wallet or keys taken—less because they'd have to be willing to cut it out of you. While some may balk at such an invasive convenience, never having to spend mental energy on remembering where you put your keys, wallet, or ID could be a game changer for executive function. For a $50 starter chip, you can check out Seattle-based maker Dangerous Things' xEM RFID chip; if you decide to go this route, I highly recommend you get it implanted by someone who has done it successfully many times before.[19] I don't have a chip yet, although I'm toying with getting one.

Motion-free neurotech

For decades, neuroengineers have been working on brain-machine interfaces that allow electrical impulses from your brain to translate directly into movements in the real world, without needing to move your hand to move a mouse or your mouth to speak to a speech-to-text program. It may sound fantastical, but technology that converts thoughts to action does exist, it just hasn't become commonplace. Typically, devices involve EEG or rely on MRI, but generally the challenge is not in reading the small, subtle electrical signals from the brain but in interpreting them correctly, quickly, and at a reasonable price point.

One company called CTRL-labs has taken a new approach. Rather than reading brain signals through the skull, they pick up electrical signals from the muscles through a device worn on the wrist. In 2018, the company was bought by Facebook for somewhere between $500 million and $1 billion.[20] This is not surprising, considering Facebook's other forays into brain-computer interfaces (BCI): they have set a goal of a "silent speech system" that could type 100 words per minute "straight from your brain."[21] We'll very likely see a version of this product available on the market within the next 5 to 10 years—or sooner.

When I'm cooking, my hands seem to get sticky instantly—which makes accessing my phone difficult. This is unfortunate, because I often need information like "Is it safe to eat pizza if you accidentally baked it with cardboard underneath?" or "How long should I boil an egg?" (As you might have guessed, I am not a great cook.) I have tried voice commands, but this has backfired in unexpected ways. I find myself yelling "Egg, not head!" and then muttering "Why would Siri think I want to know how to boil my head?" Sure, improving speech-to-text would help, but having something more contextually aware would solve the problem better.

Too bad Google Glass so fatally creeped people out; when I tried them out, I didn't mind that Google Glass "saw" what I saw. Actually, I enjoyed seeing Google Maps directions in my personal eyeglasses and felt like I'd entered a James Bond movie. Many people, however, were afraid of being recorded involuntarily through the in-Glass camera feature, and they didn't like the idea of being around "cyborgs." Many San Francisco area bars and restaurants banned Google Glass wearers from even entering their establishments.[22] Some said that Google Glass was just too early, that people would eventually get used to the technology. Compared to the "recorder pens" that automatically record lectures and even automatically take notes for students (e.g., Livescribe and other smartpen devices), Google Glass was more of a "nice to have" than a "need to have." Perhaps people were more afraid of having their faces captured than their voices, too.

SENSORY AND CREATIVITY NEUROTECH

How might you stretch your imagination if you could experience more of the world than just through your usual five senses? Neuroscientist David Eagleman devoted his popular 2015 TED Talk to a technology that could enable people to experience additional senses.[23] The versatile extra-sensory transducer (VEST) has minuscule motors that convert sounds into vibrations; the most immediate application has empowered deaf people to "hear" the world through their torsos.

With venture funding to turn the project into a wearable device,[24] Eagleman and his team have developed a wrist device for hearing-impaired people called the Buzz, which creates a sensation across the wearer's skin whenever a sound is occurring near them. This option contrasts with the cost and complexity of a cochlear implant, for instance, which also requires surgery. Eagleman and his team have already played with "sensory addition" options that could help people with ready access to their five senses add more senses. They've piped in stock market data to create a "direct perceptual experience of the economic movements on the planet," and they've also projected infrared light (something mosquitos can see),[25] ultraviolet light (which butterflies can see),[26] and Twitter data through the VEST to give a whole new way of "looking" at these data streams. Eagleman and his team have open-sourced the code behind the VEST; they hope that others will think of even more applications to further human creativity and imagination.[27]

MORAL NEUROTECH

Rather than perpetuating the biases that humans already possess, perhaps technology[28] could help us become fairer. Before a person makes a decision that is based on bias, there tend to be specific behavioral and physical indicators. Machines could record facial microexpressions, track eye movements, and detect heart rate changes that signal emotional stress—the signs that show a person is becoming confused, exhausted, or emotionally overwhelmed. The machine could then step in to prevent dangerous overreactions before they occur. In the spring of 2020, the George Floyd protests helped draw attention to racially motivated police brutality across the United States. Yet, for many, these acts came as no surprise. Researchers such as Professor Jennifer Eberhardt at Stanford have, for decades, been using neuroimaging to study how the brain processes race and produces bias. These questions—including questions of how the brain produces feelings of threat—have been the subject of a growing body of research within neuroscience. Bias has come under the microscope when it comes to gender, too. One study found that when given a choice between identical résumés with different names at the top, the résumé with a male-sounding name was consistently chosen over the identically qualified candidate with a female-sounding name.[29] What if moral neurotech could remind hiring managers to cover the names on résumés before looking at them?

Hoping technology will help us become more moral may sound like an odd suggestion, considering that, recently, AI used to make decisions in American healthcare settings,[30] hiring at a major technology company,[31] and American criminal profiling software[32] were all discovered to hold quantifiable gender, ethnic, or racial biases. In fact, the AI was detecting patterns in the data that reflect current inequities and all-too-human biases. Humans are biased and AI learns from humans, but perhaps it will ultimately be easier to correct the bias of a machine—because such corrections will be asked of a machine, not a person who could get defensive—than it is to fully eradicate bias in ourselves.

Repair neurotech

Engineering entrepreneur Elon Musk is known for boldness. Musk started out disrupting payments (PayPal), then quickly moved into solar energy (SolarCity), space travel (SpaceX), and electric cars (Tesla). Musk has long expressed concerns that AI will surpass us—and that if we can't beat it, we must join it. To do just this, he and his team at Neuralink have proposed a way to insert "a computer connection into your brain as safe and painless as LASIK eye surgery."[33]

In the summer of 2019, Musk and his team unveiled their success to date in making a "sewing machine–like" robot capable of placing tiny threads deep into the brain. Sometimes called "neural lace," this fine mesh is composed of tiny electrodes that can monitor brain function and communicate directly with machines. It could allow us to combine and amplify our intelligence with computers. Before this lofty sci-fi future comes to pass, the near-term applications of neural lace will likely be medical, like helping amputees walk or patients regain lost hearing, sight, or speech.[34]

A separate piece of science fiction that is fast becoming fact is something called neural dust. These tiny sensors, the size of grains of sand, can provide real-time data on the inner workings of the body—including reporting on the health of organs, delivering "electroceuticals" that could suppress appetite or control bladder function, or—as relates to our area of interest—read and (possibly even "write") information via electrical signals in the brain.[35] Whereas traditional electrodes need to be removed every year or two and also require holes to be drilled through the skull, neural dust would be wireless and could be injected via syringe. The dust could be "sealed in, avoiding infection and unwanted movement of the electrodes."[36]

Or, instead of implants that communicate with computers, what if we could grow entirely new brain tissue to replace parts that have become damaged or diseased? While we're still far from conducting direct transplants of brain tissue, two technologies have brought that future a bit closer. Clever use of pluripotent stem cells enabled Cambridge University neurobiologist Madeline Lancaster and her colleagues, as of a 2019 publication, to grow 3D "mini-brains" using fairly standard lab equipment. These minibrains so closely follow the ways that the human brain develops in utero that they match the shape and gene expression patterns seen throughout the first trimester of pregnancy.[37] Other scientists are working on ways to use 3D printing of specially constructed materials to make the minibrain construction process faster, more precise, and, ultimately, more repeatable.[38]

LONGEVITY NEUROTECH: COULD WE LIVE FOREVER?

As we dive deeper down the science fiction-to-fact rabbit hole, let's ask another bold question: Could we preserve our brains or minds in order to live on indefinitely?

A couple of very different approaches could get us there. The one that is doable in the short term is digital immortality. The one that is much further out is whole-brain emulation. Let's start there.

One potential path to immortality involves the billion-dollar Human Brain Project in Europe and the many research groups focused on capturing the precise wiring of the brain (in a field known as connectomics). They are aiming to simulate a brain that is essentially an amalgam of many brains, an average of sorts. To get a whole brain emulation of your particular brain, you'd probably need to die first—at least so far. At Nectome (a startup backed by seed accelerator Y Combinator), the approach to brain preservation relies on people agreeing to be pumped with embalming chemicals while they are still alive.[39] Of course, they would not do this until they were on their deathbeds and they would be under general anesthesia. This would be legal in places like California, which allows doctor-assisted suicide in cases of terminal illness, but the larger question is whether there is any scientific basis for believing that the embalming process would actually preserve enough of the connections in your brain that your memories would truly be saved. The company received almost $1 million in funding from the National Institute of Mental Health (NIMH), and as of March 2018, there was a waitlist of people interested in the service. By April 2018, however, the MIT group that had previously acted as subcontractor on the NIMH grant cut ties.[40] I reached out to the Nectome team to see if the company is still going but did not hear back.

What about the more doable approach: digital immortality?

Imagine a loved one just passed away...but you could still text with them. How? Consider the amount of conversational data you've generated through text messages, emails, social media messages, and so on. If you were to add all of that up, it's probably around a terabyte of data.[41] Next, combine that data with a machine-learning algorithm that builds a model of what you say and how you say it in different situations. Now, take that algorithm and put it inside a chatbot or digital agent that could sound like you. While it may sound sentimental, it might even be practical; for example, if we want to go into space and don't want to solve all the problems of getting our fragile human bodies to play along with zero gravity and other oddities of space travel, having a digital version of ourselves that we boot up only once we arrive at the destination might be a viable "backup" option (sorry, I couldn't resist the pun). While this remains more in the realm of academic research projects (Microsoft Research has a project on digital immortality, as does the MIT Media Lab), at least two startups have begun to tackle the problem: Replika and Eternime.[42] Going beyond Facebook pages left standing after a loved one has passed on, these startups aim to capture the digital life of a person while they live and combine it with an AI chatbot so that those who survive them can still interact with a digital version of the deceased. As AI and chatbot technology mature, I predict that more people will pursue digital immortality.

THE OPPORTUNITY OF COGNITIVE DATA

It may not sound like the sexiest topic, but to me, the most exciting area under development is cognitive data. Consider all the data you will generate in your neurohacking experiments on mood, working memory performance, attention, sleep, heart rate, what you eat, exercise, typing speed, the number of emails you send versus receive... the list goes on and on. The more experiments and tracking you do, the more data you will accumulate. The experiments we've discussed so far have been focused on helping you generate data that will allow you to make better decisions about which interventions work best for you. But as tracking improves, new and better cognitive tests come out, and interventions become more beneficial and impactful, I hope that we'll be able to build something truly revolutionary: predictive models of mental performance.

What if you could know, on any given day, what aspects of your mental performance were likely to work well and which ones less so? And the holy grail: What if you could run some baseline tests and know what types of interventions would be most likely to work well for you and which to avoid? To accomplish these feats, we'll need a lot more highly personalized data. Once we have that data, though, we'll be able to optimize our daily lives in unimaginable ways.

PERSONAL DATA NEUROTECH

We have a bit of a chicken-and-egg problem. In order to be motivated to generate your own cognitive data through neurohacking experiments, you need to be convinced that you'll gain greater understanding of your mental performance and get a better sense of what interventions will work best for you. Hopefully, you will be inspired to start running neurohacking experiments after you finish this book, and the more you run, the more data you will generate. The more data you generate, the better you'll understand yourself. The better you understand yourself, the better your predictions about the future will be. Rinse and repeat.

But imagine you had the kind of analytics and tools we have in business— say, for traders working with stock market data. Imagine a dashboard that could show you exactly how you're doing mentally at any given millisecond and even extrapolate to how you'll likely do in the future, based on your past performance. And imagine running through different scenarios, like what would happen if you tried one intervention versus another. To accomplish this, you'd eventually need tools like machine learning. Trouble is, machine learning and other statistics typically require *lots* of data to work effectively— and plenty of expertise. And of course, it couldn't just be quantity of data, it would need to be high-quality data generated by something that could measure accurately and sensitively.

Thankfully, we've got a growing precedent for big but personal data: wearable devices. As soon as we can get passive, mobile cognitive tracking that is running continuously, we'll be on our way. Imagine a wearable that doesn't just count your steps and monitor your sleep, it also tracks your cognition. Maybe it does that in an indirect biological way (for instance, eye movements and pupil size are reliable indirect indicators of attention and arousal),[43] or a direct behavioral way—you state what you're planning to do using time tracking software and it periodically asks you questions about your concentration and mood. Or perhaps it tracks the way you type, scroll, or swipe. Your typing speed and accuracy, the kinds of mistakes you tend to make, or even the particular way you press down on the keys could all be used one day as a proxy for your attention and state of mind. What I'm describing is a repurposing of recently developed software that's being used to diagnose Parkinson's early—instead of diagnosing Parkinson's, why not also predict when you're about to have a great brain day or a terrible one?[44]

What we ultimately need is just-in-time feedback, passive cognitive tracking, and customized recommendations. Frankly, this will bring on a whole new era in neurotech. But even if we have enough cognitive data, we might not want to share because we are afraid of how our "off" mental performance days could be used against us. To make it safe, these data sets will need to be as private (or as public) as each of us wants them to be. Accurate cognitive data will be extremely powerful and will make our lives better—but only if we can keep it out of the wrong hands. There are a couple of simple places to start: store any cognitive data locally (not in the cloud) and encrypt it repeatedly. I'm sure there are much better solutions than what I've outlined here and I'll be curious to see this develop.

How to Incentivize the Collection of Cognitive Data

Bryan Johnson, the founder of neurotechnology company Kernel, published a blog post in 2018 that echoed some of my own fears and hopes about greed, wisdom, and technology innovation.[45] Johnson describes two economic cycles. The first is the cycle that he fears we are currently in: a profit model of companies that prey on our attention (e.g., social media apps). Ultimately, these companies mine our mental performance for their own profit. Although some of the recent pushback to get more privacy and personal data control guidelines in place (such as the protections that the EU put in place in 2016)[46] help to some degree, the basic business model of social media and many internet companies remains the same. They capture our attention and generate data from our attention being held, and we lose control over both. Ultimately, they gain financially as we lose the portions of our day that we could be growing, learning, and advancing. Here's how Johnson illustrates that cycle:

The Economic Cycle for Human Irrelevance[47]

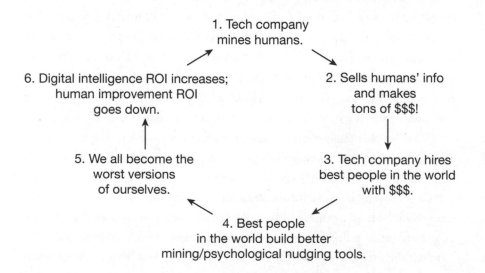

1. Tech company mines humans.

6. Digital intelligence ROI increases; human improvement ROI goes down.

2. Sells humans' info and makes tons of $$$!

5. We all become the worst versions of ourselves.

3. Tech company hires best people in the world with $$$.

4. Best people in the world build better mining/psychological nudging tools.

Thankfully, he also envisions an alternative. In this second economic cycle, *we* digitally mine ourselves and own our behaviors. When I read this, I immediately thought of the actions we take as neurohackers—we mine ourselves and our behaviors when we self-track and self-experiment. After the self-data mining step comes a step that Johnson calls "radical cognitive improvement." To a neurohacker, that would be the fruits of our labor. Johnson predicts that employers will pay upgraded employees more—perhaps through promotions, bonuses, and/or raises.

The Economic System for Radical Human Improvement[48]

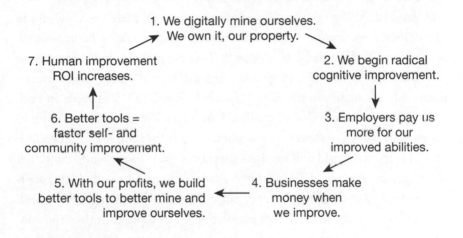

After looking at Johnson's second chart, a question came to me: What if neurohacking unlocked an economic cycle outside the employer-employee relationship? This is where gig economies could come in. The more you upgrade yourself, the more extra time you will gain. You could choose to spend that time selling your expertise on online marketplaces—such as those that already exist for businesspeople, engineers, scientists, designers, editors, and many more.[49] Or, perhaps you invent a new, useful product and sell it through an online marketplace.

While I believe that we should always own our own cognitive data, perhaps we could license it. There are already apps that pay people to share their health and lifestyle data (daily steps taken, sleep logged, meals logged, and so on) with researchers;[50] perhaps one day you will be able to share your mental performance data to facilitate the development of better cognitive interventions. My hope is that the neurohacking experiments you do will help jumpstart the cognitive data collection that will fuel a new economic system for neurohackers.

The Replication Crisis in Neuroscience

If cognitive science and neuroscience can get more money—say, from being part of a new economic cycle fueled by neurohacking and neurotechnology— I suspect that many of its scientific problems will get solved quickly. What problems? In the last few years, some of the most well-established findings in psychology have come under attack because they couldn't be replicated. Some have called it a replication crisis.[51] There are probably many reasons for this, but one may be that, in general, neuroscience and psychology experiments rely on small sample sizes—typically just 20 to 50 people in each study.[52] Sample sizes tend to be small in part because researchers aren't able to get enough money to fund larger studies. Even if the sample size problem is fixed, I suspect the field will run into another problem as the world continues to globalize; currently, 75 percent of neuroscience and psychology papers were published in the US or Europe.[53] It seems likely that many of those findings will not generalize to human populations from other countries and continents. Tara Thiagarajan, an Indian-born neuroscientist, traveled to remote villages in India and Africa and studied the brain activity of people living there.[54] Her team found that the villagers' brain dynamics were not just a little bit different from the subjects in North American and European research studies (typically college students); their neural activity was different by severalfold.[55] Research needs to be conducted not just on larger groups, but on more varied groups.

From neurodiversity and the importance of individual differences, we know that cognitive data from others won't perfectly predict your own data. To improve your cognitive performance, you'll need to do your own experiments. So, let's get you set up in your home lab—on to Part V, where you'll find your 15-minute self-experi nts!

TAKEAWAYS

1. Many technologies that used to exist solely in science fiction books and movies are coming to life.
2. Genetic selection and genetic editing may be applied to cognition. However, because intelligence and cognitive performance do not seem to be solely genetic, it's likely that genetic interventions to upgrade cognitive performance will take a while.
3. Neurotech may prevent us from making errors in executive function and emotional regulation before we make them. New technologies already promise to extend our creativity and imagination.
4. Moral neurotech could make us smarter, and more moral, too. These could decrease the role of bias in policing, hiring, in the legal system, and a variety of other social applications.
5. New technologies to repair or replace malfunctioning brain parts could allow us to extend our mental health span and cure diseases involving neurological tissue, and may even help us live forever in some capacity.
6. Cognitive data offers many opportunities. If you save the data you generate from your neurohacking experiments, you may be able to use it to predict your mental performance. We need good safety and privacy controls on who can access the data. One way is a new type of economy, one in which individuals freelance their own ideas, products, and creations on the open market.

PART V

Train and Reflect

Chapter 21

Your 15-Minute Self-Experiments

Congratulations! You made it to the self-experiments chapter. You moved through the F phase of the Focus-Select-Train-Reflect model described in chapter 4, "The Nuts and Bolts." You picked a mental target, and you gathered data on your baseline mental performance and your baseline quality of life. You've also been exposed to many interventions and hopefully picked one or two that you'd like to try out. Now, it's time to choose a self-experiment protocol, a randomization schedule, and an experimental program length. Eventually, it will be time to start buying or building your tools.

To ease your access to all of the various calendars, self-assessments, performance-based tests, and worksheets throughout the book, here are some handy shortcuts:

1. The sample self-experiment schedule is at the end of chapter 4, "The Nuts and Bolts."
2. To detect bottlenecks in your health and lifestyle, you can take the self-assessment at the end of chapter 6, "Debugging Yourself."
3. There are full self-assessments and partial performance-based tests of the four mental targets at the end of each of the chapters in Part II,

and you can find computer-based versions on my website at ericker .com. For the assessments of executive function, go to chapter 7, "The New IQ." For emotional regulation, go to chapter 8, "The New EQ." For memory and learning, go to chapter 9, "Memory and Learning." For creativity, go to chapter 10, "Creativity."

If you want to track your Life Satisfaction and Say to Do scores before, after, or during your self-experiments, you can take the survey and fill out the worksheets at the end of chapter 12, "Life Scoring."

So now let's discuss the timetable for your first self-experiment...

Choose Your Randomization Method and Experiment Duration

Once you choose your self-experiment protocol, you'll need a way to keep yourself from biasing your experiment. You won't always need to randomize, though. If you're doing an experiment that requires repeated doses to have an effect, don't randomize with any other intervention. For example, *Bacopa monnieri* takes a few weeks for its effects on learning and memory to kick in. So don't randomize with another intervention. Instead, just take *Bacopa monnieri* every day and check to see if it worked by comparing your scores as you go to your baseline and washout scores. You'll need to randomize if you're doing an a/b design where you compare the acute effects of two interventions. For example, if you're comparing how meditation versus playing Tetris affects your emotional regulation, you should randomize. Here are some options.

- One option is to alternate the intervention you use every other day. The downside of this approach is systematic bias — for example, if one of the interventions always falls on Mondays and Mondays are always very stressful for you.
- To avoid this, you can use a method that statisticians would call "sampling without replacement." Here, you never know which type of exercise you'll be doing until you're about to do it. This maintains suspense (which can be fun) and makes the experiment less biased. You can do it in the following way: Fill a bag with two different colors

of marbles. Make sure that there is an equal number of each color of marbles and that the total number of marbles matches the total number of days you'll be running your self-experiment. Each day, you'll pull out a colored marble. One color will represent one of the interventions; the other color will represent the other one. Once you've seen the color, don't put the marble back in the bag. This ensures you run the experiment for the right number of days. It has the downside of occasional "runs" where, due to chance, you end up doing the same intervention day after day.

- In a few cases, there is an option to do a "blinded" experiment. In the case of pills, say, have your neurohacking buddy randomly assign the active and placebo pills to each of the slots in a weekly pill container— make sure your buddy secretly records which pills they put into which days. Each day, take whatever pill was in the slot for that day; if you forget a day, take it as soon as you remember, but don't double up. Also, don't look at the pill as you take it, and use a drink that disguises the flavor. At the very end of the experiment, have your buddy tell you which pill you took each day.

Choose How Long You'll Run the Experiment

For most of the interventions in this book, I recommend you do each intervention type 15 to 30 times. That would mean your experiment would last between 30 and 60 days. I estimated this because I was inspired by the idea of a "just noticeable difference." In medicine, there is something called a "minimal clinically important difference."[1] Basically, scientists have found that we humans tend to notice the difference between two things only if it reaches a certain threshold of change. Let's say you held out your hands and I placed my hands on top of yours. Then, I increased the pressure on one of my hands but didn't tell you which one it was. If it was a big pressure difference, of course you would know which one it was. If it was a subtle difference, your guess would be no better than chance. The threshold when you noticed that a change occurred is the "just noticeable difference."

Like a forceful push with one hand versus a slight pressure difference, the effect of some neurohacking interventions will be large and noticeable, but others may be subtler. To detect a smaller effect, you'll need to run the intervention longer (and run the self-experiment longer overall). Based on the average size of the effects people tend to get from interventions in this book (roughly half a standard deviation difference from the pre-treatment measurement), I used computer simulations that calculated the smallest number of samples you could take and still have a reasonable chance of correctly choosing the more powerful of two interventions. That's how I got the 15 to 30 number. To be fair, this is a rough approximation—of course, the entire point of this book is that we are all different, so you should not assume that number will be exactly correct for you. Also, it's always better to take more samples than fewer (that is, run the experiment for more days) so you can be more confident that whatever results you get are not due to chance. Still, I'm impatient for answers, and I'm sure you don't have all year, either. The exception to the 15 to 30 number is cases where you're using an intervention that takes longer to take effect or has a "loading" phase, like *Bacopa monnieri*.

Choosing an Experimental Protocol

The self-experiments in this book are designed to help you compare the immediate effects of one intervention to those of another intervention (a/b tests). The mental ability tests that will measure the intervention's effects take only a minute or two, so with a 10-minute intervention, you'll still need only about 15 minutes for the whole thing each day.

	Creativity	Emotional Self-Regulation	Executive Function	Memory and Learning
Placebo	p. 257	p. 264	p. 272	p. 282
Exercise	p. 259	-	p. 272	p. 284
Light	-	p. 266	p. 275	-
Neurofeedback	-	-	p. 277	p. 285
Games	p. 261	p. 267	p. 278	p. 287
Brain Stimulation (tDCS)	p. 262	-	p. 279	-
Pills	-	p. 269	p. 281	p. 288

Once you've picked your self-experiment, you'll buy or build the tools listed in the materials list and then it will be time for you to get your baseline scores, carry out your experiment, and record your data. Don't forget to complete your washout week, in which you stop all of your interventions but keep tracking yourself by taking your daily tests. That way you'll know whether the effects of the intervention were long-lasting. Once you have your data, I'll see you in the next chapter for tips on analyzing it!

The chapter takeaways are below so that you can treat the rest of the chapter like a recipe book—feel free to browse, no need to read them all at once! To review the chapter takeaways, flip to the end of the chapter.

Introducing... the 15-Minute Self-Experiments!

You're about to embark on your self-experiments. I've picked 20 of my favorite self-experiments. They are organized by mental target. Each of the experiments has instructions on how to run the experiment, materials to make or buy, and a recommended pairing of interventions (a versus b). Each self-experiment has suggestions for ways you can customize it to make it your own, too. Finally, I've indicated the cost and complexity (potential risks, skill level required, and so on) of each self-experiment, so if you want to start with one of the easier experiments, pick one that indicates it is low in both areas.

CREATIVITY

The following experiments target creativity.

Placebos for Creativity: Scents Versus Magic Words and Scents[2]

Read more related to this self-experiment in chapter 13, "Placebo on Purpose." The experiment will compare the effects of two interventions on your creativity. In one intervention, you will take a sniff of a particular scent. The other intervention will be sniffing that scent and hearing "magic words."

Materials

- Marbles (to make the test randomized)
- A scent that makes you feel in control of your emotions (see below)
- Timer
- Tools for a creative experience (if you want to draw, write, or paint, you'll need paper or canvas, pencils or paints and brushes; if you want to solve riddles, you'll need a few riddles, etc.)

Cost

- Low ($0 to $50)

Complexity

- Low

How to Customize

- Cinnamon was used in the original study on this topic, but the following scents have been shown to boost cognition in some small studies. Feel free to use them if you want an alternative to cinnamon: lavender, lemon, orange, rosemary, and peppermint. To administer the scent, consider the following, in order from most to least recommended: scented room spray, essential oils with diffuser, sachets (with spices inside), scented candles.

Directions

1. Take the performance-based creativity test you've chosen for your daily tracking. Record your scores in your Neurohacker's Notebook.
2. Check your randomization schedule to know which intervention you are supposed to do that day.
3. On days when you are supposed to only sniff the scent, do so — but do not say the magic words.
4. On the other days sniff the scent and say these magic words: "Clinical studies have shown significant improvements through mind-body self-upgrading processes. Inhaling this scent will increase my creativity."
5. For 10 minutes, do a task that allows you to be creative. For instance, you could play with words and ideas to see if they turn into a short

story, poetry, or song, reimagine the decoration and design of your home, draw or paint anything you see or anything in your imagination, think about problems in your life and brainstorm new solutions, or solve a lateral thinking puzzle or riddle.

6. Take the creativity test from step 1 again, still smelling the scent. Record your scores.

Exercises for Creativity: Walk Versus HIIT Workout[3]

Read more related to this self-experiment in chapter 14, "Sweat." The experiment will compare the effects of two interventions on your creativity. In one intervention, you will walk. The other intervention will be a high-intensity interval training (HIIT) workout.

Materials

- Marbles (to make the test randomized)
- Timer
- Clothing you can exercise in and a space where you can lie down and kick, or walk
- If you choose to do the 7-minute HIIT workout: a chair, a surface for push-ups and sit-ups, a wall for wall-sits, and a 7-minute HIIT workout app (or just follow the instructions below)

Cost

- Low ($0 to $50)

Complexity

- Low

How to Customize

- If you have any injuries or medical concerns, check with a doctor before starting an exercise regimen. If the 7-minute HIIT workout exercises don't work for you (especially if you need lower-impact exercises), try walking or biking instead, but make sure to do HIIT with these, too — that is, one minute of high intensity, then one minute of lower intensity, then repeat. If you prefer yoga to a workout, try the sun salutation exercises.

- I like the Johnson & Johnson 7-minute HIIT workout app because it has a warmup and cooldown, too, which will push you to a little over 10 minutes total. There are plenty of other high-quality apps available. You can also use a free 7-minute workout app online and set the timer for 10 minutes so that you repeat the first 3 minutes of it at the end.
- If you decide to do any of the interventions outside in nature, you should do all of them outside. Being in nature is, itself, an intervention shown to improve mental performance.

Directions

1. Take the performance-based creativity test you've chosen for your daily tracking. Record your scores in your Neurohacker's Notebook.
2. Check your randomization schedule to know which intervention you are supposed to do that day.
3. On days when you are supposed to do the 7-minute HIIT exercise, follow the instructions below.
4. On days when you are supposed to walk, do so for 10 minutes.
5. Take the creativity test from step 1 again. Record your scores.

HIIT 7-minute workout (about 10 minutes with warmup and cooldown)[4]

Here are the instructions for the 7-minute HIIT workout. There are also some stretches before and after the workout that will take you to the 10-minute total.

1. Dynamic stretches with leg raises for 20 seconds, stretches from side to side for 20 seconds (10 on each side), torso rotations for 20 seconds (10 on each side), forearm extension stretches for 20 seconds (10 on each side).
2. Jumping jacks for 30 seconds, then rest for 10 seconds.
3. Wall sit for 30 seconds, then rest for 10 seconds.
4. Push-up for 30 seconds, then rest for 10 seconds.

5. Abdominal crunches for 30 seconds, then rest for 10 seconds.
6. Step-ups onto chair for 30 seconds, then rest for 10 seconds.
7. Squats for 30 seconds, then rest for 10 seconds.
8. Triceps dips on chair for 30 seconds, then rest for 10 seconds.
9. Plank for 30 seconds, then rest for 10 seconds.
10. High knees—running in place for 30 seconds, then rest for 10 seconds.
11. Alternating lunges for 30 seconds, then rest for 10 seconds.
12. Push-ups and rotations for 30 seconds, then rest for 10 seconds.
13. Side planks for 15 on each side (total of 30 seconds), then rest for 10 seconds.
14. Inner thigh stretches (butterfly stretches) for 30 seconds, hamstring stretches for 20 seconds (10 on each side), quadriceps stretches for 20 seconds (10 on each side), triceps stretches for 30 seconds (15 on each side).

Games for Creativity: Minecraft Versus DDR[5]

Read more related to this self-experiment in chapter 17, "Serious Games." The experiment will compare the effects of two interventions on your creativity. In one intervention, you will play Dance Dance Revolution (DDR); in the other, you will play Minecraft.

Materials

- Marbles (to make the test randomized)
- Minecraft on mobile, console, computer, or mobile devices; prices range from around $7 to $50[6]
- DDR on PlayStation, Nintendo Wii, and Xbox
- Timer

Cost

- Medium (Minecraft costs $7 to $50; DDR costs about $20)

Complexity

- Low to Medium

How to Customize

- If you don't want to try both Minecraft and DDR, you can replace either with a racing game.

Directions

1. Take the performance-based creativity test you've chosen for your daily tracking. Record your scores in your Neurohacker's Notebook.
2. Check your randomization schedule to know which intervention you are supposed to do that day.
3. On days when you are supposed to do DDR, do so for 10 minutes.
4. On days when you are supposed to play Minecraft, do so for 10 minutes.
5. Take the creativity test from step 1 again. Record your scores.

Brain Stimulation for Creativity: tDCS Versus Meditation[7]

Read more related to this self-experiment in chapter 18, "Zapping for the Better." The experiment will compare the effects of two interventions on your creativity. In one intervention, you will use tDCS (transcranial direct-current stimulation); in the other, you will do a mindfulness-based meditation.

Materials

- Marbles (to make the test randomized)
- tDCS system (device, anode, cathode, sponges, saline solution, charging cable, headband, etc.)
- Tools for a creative experience (if you want to draw, write, or paint, you'll need paper or canvas, pencils or paints and brushes; if you want to solve riddles, you'll need a few riddles, etc.)
- Timer

Cost

- High ($150+)

Complexity

- High

How to Customize

- Adjust the amount of current coming through the tDCS device so that you feel a little tingle but it still feels comfortable; some people feel a tingle right away, but some don't feel anything until they get to 2 mA. Don't exceed 2 mA, though.
- Choose a creative task that you enjoy. If you find yourself getting stressed or bored, pick a different creative task.

Directions

1. Take the performance-based creativity test you've chosen for your daily tracking. Record your scores in your Neurohacker's Notebook.
2. Check your randomization schedule to know which intervention you are supposed to do that day.
3. If you want to run a non-blinded experiment: On days when you are supposed to use tDCS, use the locations described in the paper "Non-invasive Transcranial Direct-Current Stimulation over the Left Prefrontal Cortex Facilitates Cognitive Flexibility in Tool Use."[8] Run the stimulation for 10 minutes. On days when you are supposed to do mindfulness-based meditation,[9] do so for 10 minutes. Follow the instructions below.
4. If you want to run a blinded experiment, you will need to have a friend check the randomization schedule without telling you which day it is. Then, on days they are supposed to run the "real" tDCS, they will toggle the setting to turn on real tDCS. On the other days, it should be a sham tDCS, so they will adjust that setting on the device. If your device doesn't have a sham setting, have your friend start the tDCS and then gradually turn it off over the first minute of the session.
5. In both cases, after the intervention: Choose an activity that allows you to be creative but in which you are not an expert (this tDCS protocol could suppress creativity if you are an expert in the activity) and for which you don't have to use words. For instance, you could make up tunes on a musical instrument, reimagine the decoration and design of your home, draw or paint anything you see or anything in your imagination, think about problems in your life and brainstorm

new solutions, or solve a lateral thinking puzzle or riddle. Engage in this activity for 10 minutes.

6. Take the creativity test from step 1 again. Record your scores.

Mindfulness Meditation

1. Sit in a quiet space with your eyes closed.
2. Take a deep breath. Relax.
3. With each breath, feel the air go in and out of your chest, and feel your belly contract and expand.
4. Don't control, just observe the sensations and thoughts that come and go as you sit.
5. If you notice your attention wandering, gently pull it back.
6. Counting or repeating a mantra can help you stay present.
7. If you find yourself encountering difficult feelings, memories, or sensations, try to meet them with kind, gentle, nonjudgmental care. Don't be afraid to seek out more support or resources if challenging thoughts, feelings, or experiences come up for you. If you feel you might do better with a guided meditation, there are apps for that, too! The app Headspace is one of the best-studied options[10] and provides a structured course, but other apps such as Insight Timer are free and allow for more user customization.

EMOTIONAL SELF-REGULATION

The following self-experiments target emotional regulation.

Placebos for Emotional Self-Regulation: Magic Words Versus Magic Words and Pills[11]

■ Read more related to this self-experiment in chapter 13, "Placebo on Purpose." The experiment will compare the effects of two interventions on your emotional self-regulation. In one intervention, you will take a placebo pill. In the other intervention, you will take a placebo pill and use "magic words."

Materials

- Marbles (to make the test randomized)
- Timer
- Placebo pills: order online from a branded placebo pill maker; look for white, yellow, blue, or green pills (they cost around $0.50 per pill)

Cost

- Low ($0 to $50)

Complexity

- Low
- Before you take any tests, do the following: for 5 minutes, intentionally think about something distressing to you emotionally. This could be a bad review from a boss, a fight with a loved one, fear that you won't hit a deadline, a politician doing something you feel is wrong, and so on. If you can't think of anything that makes you feel upset, tempt yourself. For example, if you struggle to control yourself when it comes to online shopping or to eating baked goods, put yourself in a situation in which you will encounter your temptation. Face your temptation and struggle to control yourself for the 5-minute period.

Directions

1. Take a placebo pill.
2. For 5 minutes, intentionally think about something distressing to you emotionally.
3. Take the performance-based emotional self-regulation test you've chosen for your daily tracking. Record your scores in your Neurohacker's Notebook.
4. Follow the instructions for a 10-minute mindfulness meditation described in the "Brain Stimulation for Creativity" self-experiment on page 264.
5. Check your randomization schedule to know which intervention you are supposed to do that day.
6. On the days when you are supposed to say magic words, say the following: "Clinical studies have shown significant improvement through

mind-body self-upgrading processes. Taking these pills will help me regulate my emotions."

7. Take the emotional self-regulation test from step 1 again. Record your scores.

8. Note: If you're not feeling better by the end, do something self-soothing (e.g., listen to calming music, go for a walk, talk to a friend, etc.).

Light for Emotional Self-Regulation: Blue Versus Ambient Light[12]

Read more related to this self-experiment in chapter 15, "Let There Be (Blue) Light." The experiment will compare the effects of two interventions on your emotional regulation. In one intervention, you will use blue light. The other intervention will be to use whatever light is currently available in your environment (no blue light).

Materials

- Marbles (to make the test randomized)
- Blue light: The Philips goLITE BLU is the product tested in two of the research studies cited in chapter 15. Typically, blue lights cost around $80 online.

Cost

- Low to Medium ($0 to $150)

Complexity

- Low

How to Customize

- Some research shows similar effects on mood from very white lights. Generic natural-spectrum light therapy lamps cost less; I ordered the Verilux Happylight Compact Personal for around $40.
- Adjust the light intensity level for comfort and so you don't get eye strain.
- The purpose of thinking about something distressing is to get you into an emotional state that you will then have to regulate. This could

be a bad review from a boss, a fight with a loved one, fear that you won't hit a deadline, a politician doing something you feel is wrong, and so on. If you can't think of anything that makes you feel upset, tempt yourself. For example, if you struggle to control yourself when it comes to online shopping or to eating baked goods, put yourself in a situation in which you will encounter your temptation. Face your temptation and struggle to control yourself for the 5-minute period.

Directions

1. For 5 minutes, intentionally think about something distressing to you emotionally.
2. Take the performance-based emotional self-regulation test you've chosen for your daily tracking. Record your scores in your Neurohacker's Notebook.
3. Check your randomization schedule to know which intervention you are supposed to do that day.
4. On days when you are supposed to use blue light, turn it on for 10 minutes. On days when you are supposed to use ambient light, do so for 10 minutes (that is, don't turn on the blue light).
5. Take the emotional self-regulation test from step 1 again. Record your scores.

Games for Emotional Self-Regulation: Tetris Versus Meditation[13]

Read more related to this self-experiment in chapter 17, "Serious Games." The experiment will compare the effects of two interventions on your emotional regulation. In one intervention, you will play Tetris; in the other, you will write a diary entry.

Materials

- Marbles (to make the test randomized)
- Tetris (free)
- Writing tools (paper and pencil or a computer)

Cost

- Low ($0)

Complexity

- Low

How to Customize

- If neither Tetris nor writing in a diary appeals to you, consider one of the following alternative interventions: Nevermind for adults, Mightier for children, or SuperBetter.
- The purpose of thinking about something distressing is to get you into an emotional state that you will then have to regulate. This could be a bad review from a boss, a fight with a loved one, fear that you won't hit a deadline, a politician doing something you feel is wrong, and so on. If you can't think of anything that makes you feel upset, tempt yourself. For example, if you struggle to control yourself when it comes to online shopping or to eating baked goods, put yourself in a situation in which you will encounter your temptation. Face your temptation and struggle to control yourself for the 5-minute period.

Directions

1. For 5 minutes, intentionally think about something distressing to you emotionally.
2. Take the performance-based emotional self-regulation test you've chosen for your daily tracking. Record your scores in your Neurohacker's Notebook.
3. Check your randomization schedule to know which intervention you are supposed to do that day.
4. On days when you are supposed to play Tetris, do so for 10 minutes.
5. On days when you are supposed to meditate, follow the instructions for a 10-minute mindfulness meditation described in the "Brain Stimulation for Creativity" self-experiment on page 264.
6. Take the emotional self-regulation test from step 1 again. Record your scores.

Pills for Emotional Self-Regulation: Kava Versus Herbal Tea[14]

Read more related to this self-experiment in chapter 19, "A Pill a Day." The experiment will compare the effects of two interventions on your emotional regulation. In one intervention, you will drink kava; in the other, you will drink an herbal tea of your choosing.

Materials

- Marbles (to make the test randomized)
- Herbal tea of your choice (preferably one that helps you feel emotionally balanced)
- If you choose kava in drink form: Buy certified "noble" kava that comes from the root (not the rest of the plant). Look for vendors that get their kava regularly tested by third-party labs.[15] If you buy it in a kava lounge, it will cost $5 to $8.[16] If you order it online and prepare it yourself, it will cost $4 or $5 per serving and can come in multiserving packages such that you'll be paying around $50 per package.[17] To disguise the earthy taste of the kava (so you can run a properly blinded experiment), combine coconut milk and tropical juice with the kava. Then, for your control group, use just the coconut milk and tropical juice.[18] Another disguise you could try is cocoa powder; you can create a chocolate milk kava smoothie by combining it with coconut milk and other ingredients.[19] You'll need a friend to help you truly mask your experiment.
- If you choose kava in pill form: Research studies tend to focus on the extracts known as WS1490 or LI 150. However, these are hard to get for many consumers. For pill forms of kava, you'll find dose recommendations on Examine.com's page about kava: https://examine.com/supplements/kava/#how-to-take.
- Caution: Unlike with many other substances, you don't need more to get the same effect you get the very first time—instead, kava seems to become more powerful the more frequently you take it. So, don't overdo it—a little goes a long way. If you want to consume kava in drink form, I wouldn't advise having more than one shell at a time. Bring a designated driver, and be very careful about how often you

take it (although it tends to have a reverse tolerance, so it actually affects you more the more you take it). There are more recommendations in chapter 19. On the weekend, you could find a kava bar as a more intellectually engaging alternative to the standard bar and run your experiments with some friends. Don't operate machinery or drink alcohol at the same time as taking kava, and if you have any liver issues, avoid it altogether.

Cost

■ Low ($0 to $50)

Complexity

■ Medium

How to Customize

■ The purpose of thinking about something distressing is to get you into an emotional state that you will then have to regulate. This could be a bad review from a boss, a fight with a loved one, fear that you won't hit a deadline, a politician doing something you feel is wrong, and so on. If you can't think of anything that makes you feel upset, tempt yourself. For example, if you struggle to control yourself when it comes to online shopping or to eating baked goods, put yourself in a situation in which you will encounter your temptation. Face your temptation and struggle to control yourself for the 5-minute period.

If you want to make it a double-blind experiment, you'll need a friend to make up capsules for you, microcrystalline cellulose powder, kava powder, and empty capsules. Your friend should make an equal number of pills containing cellulose powder (your placebo option, in place of the herbal tea) and pills containing kava powder, fill one container with active pills and one container with inactive pills, and label them to match the colors of the marbles (e.g., "red" and "blue"), so when you pull a marble of that color you'll know which container to take a pill from.

Directions

1. For 5 minutes, intentionally think about something distressing to you emotionally.

2. Take the performance-based emotional self-regulation test you've chosen for your daily tracking. Record your scores in your Neurohacker's Notebook.

3. Check your randomization schedule to know which intervention you are supposed to do that day.

4. Drink the herbal tea or take the kava. Kava takes 15 to 30 minutes to affect you, so wait that long before taking the second emotional regulation test. During that time, do the mindfulness meditation described in the "Brain Stimulation for Creativity" experiment on page 264.

5. Take the emotional regulation test from step 1 again. Record your scores.

EXECUTIVE FUNCTION

The following self-experiments target executive function.

Placebos for Executive Function: Magic Words Versus Magic Words and Objects[20]

Read more related to this self-experiment in chapter 13, "Placebo on Purpose." The experiment will compare the effects of two interventions on your executive function. In one intervention, you will wear a white lab coat and use "magic words." The other intervention will be the same except no "magic words."

Materials

- Marbles (to make the test randomized)
- White lab coat
- Timer
- Writing tools (paper and pencil or a computer)

Cost

- Low ($0 to $50)

Complexity

- Low

How to Customize

- Instead of a white lab coat, feel free to personalize: Is there a profession that seems sharper than a scientist or doctor to you? If that profession has a recognizable uniform or prop (such as a fountain pen, calculator, stethoscope, magic wand, glasses, etc.), wear it / hold it instead of the white lab coat.

Directions

- Take the performance-based executive function test you've chosen for your daily tracking. Record your scores in your Neurohacker's Notebook.
- Check your randomization schedule to know which intervention you are supposed to do that day.
- Put on your lab coat.
- On days when you are supposed to, say the following magic words: "Clinical studies have shown significant improvement through mind-body self-upgrading processes. Wearing this lab coat will increase my executive function."
- Do a 15-minute task that involves executive function (for example, assessing how yesterday went and planning today using Say to Do methods from chapter 12). At the end, record in your Neurohacker's Notebook how well you feel your executive function task went (on a scale from 1 to 5, with 1 being poor, 3 being medium, 5 being perfect).
- Take the executive function test from step 1 again, still wearing the lab coat. Record your scores in your Neurohacker's Notebook.

Exercise for Executive Function: Coordination Exercises or HIIT Versus No Exercise[21]

Read more related to this self-experiment in chapter 14, "Sweat." The experiment will compare the effects of two interventions on your executive function. In one intervention, you will do coordination exercises or HIIT. The other intervention will be to do no exercise.

Materials

- Marbles (to make the test randomized)
- Timer
- Clothing you can exercise in and a space where you can lie down and kick, or walk
- If you choose to do the 7-minute HIIT workout: a chair, a surface for push-ups and sit-ups, a wall for wall-sits, and a 7-minute HIIT work-out app (or just follow the instructions below)

Cost

- Low ($0 to $50)

Complexity

- Medium

How to Customize

- If you have any injuries or medical concerns, check with a doctor before starting an exercise regimen. If the tae kwon do workout or the 7-minute HIIT workout exercises don't work for you (especially if you need lower-impact exercises), try walking or biking instead, but make sure to do HIIT with these, too—that is, one minute of high intensity, then one minute of lower intensity, then repeat). Or, if you prefer yoga to a workout, try the sun salutation exercises.
- You have a choice of three exercise protocols: (1) a 7-minute HIIT workout, (2) a tae kwon do workout, or (3) a yoga sequence. You may want to watch the videos on my website to get the movements right or you can just follow the instructions below.
- If you choose the HIIT workout, I like the Johnson & Johnson 7-minute HIIT workout app because it has a warmup and cooldown, too, which will push you to a little over 10 minutes total. There are plenty of other high-quality apps available. You can also use a free 7-minute workout app online and set the timer for 10 minutes so that you repeat the first 3 minutes of it at the end.
- If you decide to do any of the interventions outside in nature, you should do all of them outside. Being in nature is, itself, an intervention shown to improve mental performance.[22]

Directions

1. Take the performance-based executive function test you've chosen for your daily tracking. Record your scores in your Neurohacker's Notebook.
2. Check your randomization schedule to know which intervention you are supposed to do that day.
3. On days when you are supposed to exercise, follow the instructions for your chosen 10-minute exercise (see below).
4. On days when you are supposed to do nothing, sit and rest for 10 minutes.
5. Take the executive function test from step 1 again. Record your scores.

Exercise Instructions

HIIT 7-minute workout: See the instructions in the "Exercises for Creativity" experiment on pages 260–261.

Tae kwon do workout: (take approximately 43 seconds for each of the 14 movements below).[23]

1. Rotate the shoulders, arms, hips.
2. Rotate the knees, lift the knees and pull your knees outward and inward.
3. Light jogging, followed by star jumps.
4. Push-ups then sit-ups.
5. Light jogging, followed by star jumps.
6. Leg raising, right and left.
7. Switch stance, then double switch.
8. Turning kicks.
9. Axe kicks.
10. Front kicks.
11. Side kicks.
12. Back kicks or reverse side kicks.
13. Reverse turning kicks.
14. Squats and side stretching.

Yoga (sun salutation sequence: inhale and exhale with the instructions below; take 50 seconds for each of the 12 poses):[24]

1. Prayer pose — exhale
2. Raised arm salute — inhale
3. Forward fold / hand to foot — exhale
4. Left leg lunge — inhale
5. Downward dog — exhale
6. Staff pose / 8-point salute — inhale
7. Cobra — exhale
8. Downward dog — inhale
9. Right leg lunge — exhale
10. Forward fold / hand to foot — inhale
11. Raised arm salute — exhale
12. Prayer pose — inhale

Light for Executive Function: Blue Light Versus Caffeine[25]

Read more related to this self-experiment in chapter 15, "Let There Be (Blue) Light." The experiment will compare the effects of two interventions on your emotional regulation. In one intervention, you will use blue light; in the other, you will use caffeine.

Materials

- Marbles (to make the test randomized)
- Blue light: The Philips goLITE Blu is the product tested in two of the research studies cited in chapter 15. Typically, blue lights cost around $80 online.
- 40 mg of caffeine (e.g., a cup of black tea, half a cup of coffee, or a pill)

Cost

- Low to Medium ($0 to $150)

Complexity

- Low

How to Customize

- Some research shows similar effects on mood from very white lights; it's possible that white lights could provide benefits to executive function, too. Generic natural spectrum light therapy lamps cost less; I ordered the Verilux Happylight Compact Personal for around $40.
- Adjust the light intensity level for comfort and so you don't get eye strain.
- Pick a form of caffeine (tea, coffee, or pill) that you like but don't love in order to control your own bias.

Directions

1. Take the performance-based executive function test you've chosen for your daily tracking. Record your scores in your Neurohacker's Notebook.
2. Check your randomization schedule to know which intervention you are supposed to do that day.
3. On days when you are supposed to use blue light, turn it on for 10 minutes. Use the time to do the following task, which involves executive function: Reflect on yesterday and plan your coming day. Use the Say to Do methods from chapter 12. At the end, record in your Neurohacker's Notebook how well you feel this executive function task went (on a scale from 1 to 5, with 1 being poor, 3 being medium, 5 being perfect).
4. On days when you are supposed to take caffeine, do so. Go do something else for 30 to 60 minutes, as that's how long it takes for caffeine to hit maximum efficacy (although individual variation can be significant). Then do the following task, which involves executive function: For 10 minutes, reflect on yesterday and plan your coming day. Use the Say to Do methods from chapter 12. At the end, record in your Neurohacker's Notebook how well you feel this executive function task went (on a scale from 1 to 5, with 1 being poor, 3 being medium, 5 being perfect).
5. Take the executive function test from step 1 again. Record your scores.

Neurofeedback for Executive Function: Neurofeedback Versus Mindfulness Meditation[26]

Read more related to this self-experiment in chapter 16, "Rewriting the Brain's Signature." The experiment will compare the effects of two interventions on your emotional regulation. In one intervention, you will do neurofeedback; in the other, mindfulness meditation.

Materials

- Neurofeedback headset ($150 to $300) and app (usually free), or appointments with neurofeedback clinicians to use the professional equipment in their office. Personally, I like using the Muse headset (made by the Canadian company Interaxon) and associated app for meditation; I've even used it for research. This headset is also useful if you want to train remotely with a clinician through the Israeli startup Myndlift. Therapists' hourly rates are typically based on the local cost of living and doing business and on their credentials; someone with a family and marriage therapy degree might charge less, but a PhD who includes neurofeedback as one of many services might charge over $100/hour. There's more information here, too: https://www.aapb.org/i4a/pages/index.cfm?pageid=3387.
- 10-minute guided meditation (free on YouTube or an app)

Cost

- High ($150+)

Complexity

- Medium

How to Customize

- Decide whether you'd rather work with a professional clinician in their office, use the programs on the app that comes with a consumer device at home, or do a combination where you work with a remote clinician over the internet but use a consumer device at home. If you decide to use a consumer device, get a device with an adjustable headband; people's heads vary dramatically in size.

- If you have a medical concern and/or you want a more customized approach and are willing to pay more and spend more time, you can work one-on-one with a professional neurofeedback practitioner. For a BCIA-certified neurofeedback therapist in your area, check out https://certify.bcia.org/4dcgi/resctr/search.html.

Directions

1. Take the performance-based executive function you've chosen for your daily tracking. Record your scores in your Neurohacker's Notebook.
2. Check your randomization schedule to know which intervention you are supposed to do that day.
3. On days when you are supposed to do neurofeedback, do so for 10 minutes.
4. On days when you are supposed to do mindfulness meditation, do so for 10 minutes. Follow the instructions under the "Brain Stimulation for Creativity" experiment on page 264.
5. Take the executive function test from step 1 again. Record your scores.

Games for Executive Function: Brain Games Versus Brain Games[27]

Read more related to this self-experiment in chapter 17, "Serious Games." The experiment will compare the effects of two interventions on your executive function. In one intervention, you will play BrainHQ; in the other, dual n-back. Note that you'll need to choose a different performance-based test if you choose dual n-back as your main intervention.

Materials

- Marbles (to make the test randomized)

Cost:

- $0 to $50 (there are free versions of both BrainHQ and the dual n-back, but for more features, you'll need to pay)

Complexity

- Low

How to Customize

- For best results, pick the game options that you're most interested in.
- Alternatively, you could compare the effects of playing Portal 2 to Lumosity. Or, you could compare any of them to getting as far as you can in a crossword puzzle in 10 minutes.

Directions

1. Take the performance-based executive function test you've chosen for your daily tracking. Record your scores in your Neurohacker's Notebook.
2. Check your randomization schedule to know which intervention you are supposed to do that day.
3. On days when you are supposed to play BrainHQ, do so for 10 minutes.
4. On days when you are supposed to play dual n-back, do so for 10 minutes.
5. Take the executive function test from step 1 again. Record your scores.

Brain Stimulation for Executive Function: tDCS Versus Meditation[28]

Read more related to this self-experiment in chapter 18, "Zapping for the Better." The experiment will compare the effects of two interventions on executive function tests. For this self-experiment, use the book n-back (see pages 76–78) or a computer-based n-back test, as other types of executive function test are less likely to show a change. In one intervention, you will use tDCS (transcranial direct-current stimulation); in the other, you will do a mindfulness-based meditation.

Materials

- Marbles (to make the test randomized)
- tDCS system (device, anode, cathode, sponges, saline solution, charging cable, headband, etc.)
- Writing tools (paper and pencil or a computer)
- Timer

Cost

- High ($150+)

Complexity

- High

How to Customize

- Adjust the amount of current coming through the tDCS device so that you feel a little tingle but it still feels comfortable; some people feel a tingle right away, but some don't feel anything until they get to 2 mA. Don't exceed 2 mA, though.

Directions

1. Take the performance-based executive function test you've chosen for your daily tracking. Record your scores in your Neurohacker's Notebook.
2. Check your randomization schedule to know which intervention you are supposed to do that day.
3. If you want to run a non-blinded experiment: On days when you are supposed to use tDCS, use the locations described in the paper "Effects of Prefrontal tDCS on Executive Function: Methodological Considerations Revealed by Meta-Analysis."[29] Run the stimulation for 10 minutes.
4. If you want to run a blinded experiment, you will need to have a friend check the randomization schedule without telling you which day it is. Then, on days they are supposed to run the "real" tDCS, they will toggle the setting to turn on real tDCS. On the other days, it should be a sham tDCS, so they will adjust that setting on the device. If your device doesn't have a sham setting, have your friend start the tDCS and then gradually turn it off over the first minute of the session.
5. In both cases, after the intervention: Reflect on yesterday and plan your coming day. Use the Say to Do methods from chapter 12. At the end, record in your Neurohacker's Notebook how well you feel this 10-minute executive function task went (on a scale from 1 to 5, with 1 being poor, 3 being medium, 5 being perfect).
6. Take the executive function test from step 1 again. Record your scores.

Pills for Executive Function: Caffeine + L-Theanine Versus Placebo Pills[30]

Read more related to this self-experiment in chapter 19, "A Pill a Day." The experiment will compare the effects of two interventions on your executive function. In one intervention, you will take caffeine and L-theanine; in the other, placebo pills.

Materials

- Marbles (to make the test randomized)
- You can buy a combination pill that has 100 to 200 mg of caffeine and 100 to 200 mg of L-theanine. Combined caffeine and L-theanine pills cost $0.15 to $0.40 per pill online. Go for a reputable brand with good reviews and third-party testing; it's worth paying a little extra to avoid toxins.
- Placebo pills: order online from a branded placebo pill maker (they cost around $0.50 per pill).
- If you want to make it a double-blind experiment, you'll need a friend to make up capsules for you, microcrystalline cellulose powder, caffeine powder, L-theanine powder, and empty capsules. Your friend should make an equal number of pills containing cellulose powder (your placebo option) and pills containing equal parts coffee and L-theanine powder, fill one container with active pills and one container with inactive pills, and label them to match the colors of the marbles (e.g., "red" and "blue"), so when you pull a marble of that color you'll know which container to take a pill from.

Cost

- Low ($0 to $50)

Complexity

- Low

How to Customize

- Especially if you think you'll be sensitive to caffeine, start with 100 mg caffeine. For the average adult, the max caffeine you should consume per day is around 400 mg per day. If you're pregnant or breastfeeding, the American College of Obstetricians and Gynecologists recommends that you max out at 100 to 200 mg of caffeine per day (the lower end is for pregnancy and the higher end is for breastfeeding).

Directions

1. Take the performance-based executive function test you've chosen for your daily tracking. Record your scores in your Neurohacker's Notebook.
2. Check your randomization schedule to know which intervention you are supposed to do that day.
3. Take one pill. Go do something else for 30 to 60 minutes, as that's how long it takes for caffeine to hit maximum efficacy (although individual variation can be significant). Then do the following task, which involves executive function: For 10 minutes, reflect on yesterday and plan your coming day. Use the Say to Do methods from chapter 12. At the end, record in your Neurohacker's Notebook how well you feel this executive function task went (on a scale from 1 to 5, with 1 being poor, 3 being medium, 5 being perfect).
4. Take the executive function test from step 1 again. Record your scores.

LEARNING AND MEMORY

The following self-experiments target learning and memory.

Placebos for Learning and Memory: Magic Words Versus Visualization[31]

Read more related to this self-experiment in chapter 13, "Placebo on Purpose." The experiment will compare the effects of two interventions on your learning and memory. In one intervention, you will learn using visualization and "magic words"; in the other, you will learn by practicing physically (the usual way).

Materials

■ Marbles (to make the test randomized)

Cost

■ Low ($0 to $50)

Complexity

■ Low

How to Customize

■ Pick a subject to learn that you are passionate about. For this self-experiment, it helps to choose a physical skill, like playing the guitar or piano or learning a new dance.

■ An alternative to the randomization schedule is that you could do all visualization for one week and then all nonvisualization (physical practice) for the next week. If you run the experiment this way, make sure to do your pre-test and post-test at the beginning and end of each week so you can compare the performances from each method. Whichever method you choose, you'll still need to repeat this for enough weeks that you end up with 15 to 30 sessions of each type of intervention.

Directions

1. Take the performance-based learning and memory test you've chosen for your daily tracking. Record your scores in your Neurohacker's Notebook.
2. Check your randomization schedule to know which intervention you are supposed to do that day.
3. On days when you are supposed to do visualization, begin by saying these "magic words": "Clinical studies have shown significant improvement through mind-body self-upgrading processes. Visualization will increase my learning and memory." Then practice the skill without moving your body. Just close your eyes, remain motionless, and imagine doing the activity for 10 minutes.

4. On days when you are not supposed to do visualization, just practice the skill physically, as you normally would, for 10 minutes.

5. Take the learning and memory test from step 1 again. Record your scores.

Exercise for Learning and Memory: HIIT Versus Steady Exercise[32]

Read more related to this self-experiment in chapter 14, "Sweat." The experiment will compare the effects of two interventions on your learning and memory. In one intervention, you will do a HIIT (high-intensity interval training) workout; in the other you will exercise at a steady rate.

Materials

- Marbles (to make the test randomized)
- Timer
- Clothing you can exercise in and a space where you can lie down and kick
- If you choose to do the 7-minute HIIT workout: a chair, a surface for push-ups and sit-ups, a wall for wall-sits, and a 7-minute HIIT workout app

Cost

- Low ($0 to $50)

Complexity

- Low

How to Customize

- Pick a subject to learn that you are passionate about.
- If you have any injuries or medical concerns, check with a doctor before starting an exercise regimen. For a HIIT workout, determine how fast you should go in your "on" interval and how slow you should go in your "off" interval.

- I like the Johnson & Johnson 7-minute HIIT workout app because it has a warmup and cooldown, too, which will push you to a little over 10 minutes total. There are plenty of other high-quality apps available. You can also use a free 7-minute workout app online and set the timer for 10 minutes so that you repeat the first 3 minutes of it at the end.
- Pick a type of steady exercise that works for your current fitness and interest level: biking, walking, or running.
- If you decide to do any of the interventions outside in nature, you should do all of them outside. Being in nature is, itself, an intervention shown to improve mental performance.

Directions

1. Take the performance-based learning and memory test you've chosen for your daily tracking. Record your scores in your Neurohacker's Notebook.
2. Check your randomization schedule to know which intervention you are supposed to do that day.
3. On days when you are supposed to do the 7-minute HIIT workout, follow the instructions in the "Exercises for Creativity" self-experiment on pages 260–261.
4. On days when you are supposed to exercise at a steady rate, bike, walk, or run at a low to moderate pace (you should be able to keep up a conversation easily) for 10 minutes.
5. Take the learning and memory test from step 1 again. Record your scores.

Neurofeedback for Learning and Memory: Neurofeedback Versus Meditation[33]

Read more related to this self-experiment in chapter 16, "Rewriting the Brain's Signature." The experiment will compare the effects of two interventions on your emotional regulation. In one intervention, you will do neurofeedback; in the other, mindfulness meditation.

Materials

- Marbles (to make the test randomized)
- Neurofeedback headset ($150 to $300) and app (usually free), or appointments with neurofeedback clinicians to use the professional equipment in their office. Therapists' hourly rates are typically based on the local cost of living and doing business and on their credentials; someone with a family and marriage therapy degree might charge less, but a PhD who includes neurofeedback as one of many services might charge over $100/hour. There's more information here, too: https://www.aapb.org/i4a/pages/index.cfm?pageid=3387.
- 10-minute guided meditation (free on YouTube or an app)

Cost

- High ($150+)

Complexity

- Medium

How to Customize

- Decide whether you'd rather work with a professional clinician in their office, use the programs on the app that comes with a consumer device at home, or do a combination where you work with a remote clinician over the internet but use a consumer device at home. If you decide to use a consumer device, get a device with an adjustable head-band; people's heads vary dramatically in size.
- If you have a medical concern and/or you want a more customized approach and are willing to pay more and spend more time, you can work one-on-one with a professional neurofeedback practitioner. For a BCIA-certified neurofeedback therapist in your area, check out https://certify.bcia.org/4dcgi/resctr/search.html.

Directions

1. Take the performance-based learning and memory test you've chosen for your daily tracking. Record your scores in your Neurohacker's Notebook.

2. Check your randomization schedule to know which intervention you are supposed to do that day.

3. On days when you are supposed to do neurofeedback, do so for 10 minutes.

4. On days when you are supposed to do mindfulness meditation, do so for 10 minutes. Follow the instructions under the "Brain Stimulation for Creativity" self-experiment on page 264.

5. Take the learning and memory test from step 1 again. Record your scores.

Games for Learning and Memory: Spaced Repetition Versus Typical Study Methods[34]

Read more related to this self-experiment in chapter 17, "Serious Games." The experiment will compare the effects of two interventions on your learning and memory. In one intervention, you will use spaced repetition software to learn new material; in the other, you will use a typical learning method (physical flash cards). Because these interventions are learning aids—they affect how you learn the material while you use the aid but not how you learn material generally—you will be testing yourself on the material learned, not on your learning and memory abilities per se.

Materials

- Spaced repetition app
- Tools to make flash cards (paper, markers)

Cost

- Low ($0 to $50)

Complexity

- Low

How to Customize

- Pick a spaced repetition app at a price point and of a style that fits you best. Anki is one of the best known and it is free, but there are plenty from which to choose.

- Pick a topic to learn that you're excited to master and that involves a lot of new words or concepts (for example, a foreign language or an area of science). There should also be at least eight tests on the subject matter available so that you can assess your learning as you go. Split the words or concepts you plan to learn in half. Half you will learn using spaced repetition and half you will learn using your typical study method.

Directions

1. Take a baseline test to assess your knowledge of the topic you plan to learn.
2. Flip a coin to decide whether you will start with spaced repetition or your typical study method.
3. Begin studying using whichever intervention the coin flip indicated. Each day for a week, spend 15 minutes learning using that intervention. Record how many words you studied each day in your Neurohacker's Notebook. At the end of the week, take a test to assess how much you learned that week.
4. Switch to the other intervention for the next week. Alternate each week for at least eight weeks.
5. At the end of your self-experiment, compare how many words you learned each day using the two interventions. Also compare how well you did on your end-of-week tests. Which intervention led to better learning for you?

Pills for Learning and Memory: *Bacopa monnieri* Versus Placebo[35]

Read more related to this self-experiment in chapter 19, "A Pill a Day." The experiment will compare the effects of two interventions on your executive function. In one intervention, you will take *Bacopa monnieri*. The other intervention will be to take placebo pills.

Materials

- Marbles (to make the test randomized)

- For dosage recommendations, refer to Examine.com's page on *Bacopa monnieri:* https://examine.com/supplements/bacopa-monnieri/#how -to-take.
- A fatty food or drink
- Placebo pills: order online from a branded placebo pill maker (they cost around $0.50 per pill)
- If you want to make it a double-blind experiment, you'll need a friend to make up capsules for you, microcrystalline cellulose powder, *Bacopa monnieri* powder, and empty capsules. Your friend should make an equal number of pills containing cellulose powder (your placebo option) and pills containing *Bacopa monnieri* powder, fill one container with active pills and one container with inactive pills, and label them to match the colors of the marbles (e.g., "red" and "blue"), so when you pull a marble of that color you'll know which container to take a pill from.

Cost

- Low ($0 to $50)

Complexity

- Low

How to Customize

- Some neurohackers report nausea, cramping, bloating, and diarrhea after taking *Bacopa monnieri,* so definitely take it with a meal that contains some fat so it can be absorbed properly (hence the traditional use of ghee).[36] Many people report feeling very relaxed and even a bit sleepy after taking it, so you could try taking it at night.[37]

Directions

1. Take the performance-based learning and memory test you've chosen for your daily tracking. Record your scores in your Neurohacker's Notebook.
2. No need to randomize; I wouldn't recommend you do an a/b test design. Just rely on a comparison with your baseline period and washout to know whether the intervention worked.

3. Studies show an almost immediate decrease in anxiety levels after taking *Bacopa monnieri,* but the improvements to memory didn't seem to show up until after participants had taken it for 12 weeks.[38] For that reason, you'll probably want to keep taking *Bacopa monnieri* for 12 weeks to test whether the learning and memory effects occur for you.

TAKEAWAYS

1. You confirmed your choices of mental target and intervention.
2. You chose how to randomize (for example, alternating days or sampling without replacement) and how long to run your self-experiment (for example, 15 to 30 sessions).
3. You got to choose from 20 different experimental protocols.

Chapter 22

Pretty Pictures

> **Time Investment:** 6 minutes
> **Goal:** To analyze the data your experiment generated and consider your next move

You've run your self-experiment. You've gathered your data. Perhaps you feel like you improved your mental performance, but you want to run the numbers to see how it really worked. How do you analyze all the data you've gathered? Two ways may prove useful to you: a statistical way and a graphical way.

Let's start with the graphical way. There's some truth behind the expression "a picture is worth a thousand words." The amount of your brain devoted to making sense of numbers is relatively small and recently evolved compared to the part devoted to visual processing.[1] It also takes your brain far longer to process symbolic information—like text—than it does visual information.[2] Many people feel uncomfortable making judgments based on numbers, but most of us feel comfortable making judgments about pictures. That's right, I want you to draw some pretty pictures and forget about statistics for now.

You'll be looking for three things in your charts: the trend, the wobble, and the level.[3] The trend is whether the data points seem to go up or down overall. The wobble is how tightly clustered the points are to each other. The level is the rough average of all the points, if you had to draw a straight line through them. First, make a scatter plot graph with days on the *x*-axis and test performance on the *y*-axis. Any spreadsheet program will do. Color the data points from the baseline period, intervention period, and washout period differently—say, red, green, and blue—or use different symbols for each period. Next, look at the graphs across the three phases: Did your performance seem to improve over time, and then hit a plateau during the baseline period? Then did it start to rise again during your intervention period? And then stay roughly the same during the washout? That would be a pretty promising picture. This assumes you chose an experimental design with just one intervention (such as *Bacopa monnieri,* a nootropic where it takes a few weeks for its learning and memory effects to kick in). How would you assess whether your intervention was really improving your mental performance: Would you take the average score on your mental performance tests? Not so fast.

Let's discuss averages for a moment. It is very easy to get fooled by them, which is why I've so strongly encouraged you to draw out your data in graphs first. To illustrate why averages can be misleading and how you can learn more from analyzing the graph of the entire experiment, let's look at a set of examples that I've dubbed, rather modestly, "Ricker's Quintet."[4] They show graphs of data from sample self-experiments. Which of these do you think the intervention worked, which didn't, and which were just inconclusive?

Ricker's Quintet: Figures 1–4

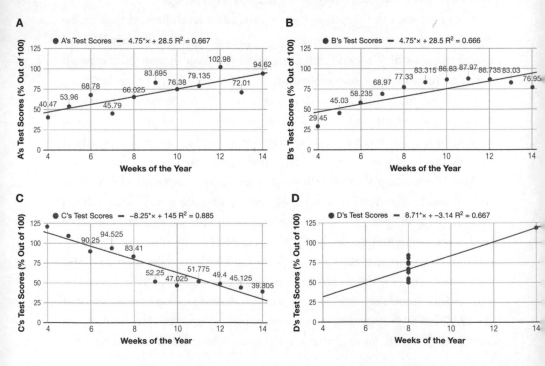

Ricker's Quintet: Figure 5

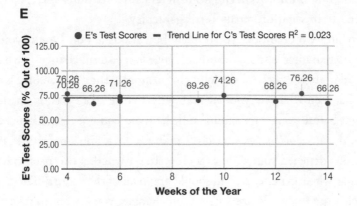

You're on the right track if your answer was something along the lines of: A is trending up, so the intervention might be helping. B initially trends up, but then it starts to fall back down after week 10, so that intervention might have petered out. C looks like it's trending down, so if anything, the intervention seems to be making the person worse. D is inconclusive, because most of the data points were taken on the same day—the first day. Even though the final score is much higher, there was too long a lag in between, so there could have been other things that caused the gain, not necessarily the intervention.

These graphs share something in common. You may have guessed already, but here's the kicker: they all share exactly the same average: 71.25 percent.

The reason I included a bunch of dissimilar-looking charts with the same average is to underscore a single point: if you had just looked at their averages, you would have thought they were all similar experiments. However, it's clear that A and B are both moderately promising interventions, but the others are not. Don't get fooled by averages. Look at a scatter plot of your data before you start computing averages or any other statistics. Otherwise, you'll overlook important trends. The trends are often the real story.

Now let's look at just the last image, figure E. Would you conclude that the intervention did anything for you? Without getting into any statistics, just from observing how most of the upticks are balanced by similar-sized downticks, I would guess that this intervention did little in this self-experiment. If you see something like this after your self-experiment, you may want to assume that the interventions didn't affect you very much.

Now let's look at how to interpret an a/b experiment. This is the type where you randomize between two different interventions—say, comparing the effects of meditation versus Tetris gameplay.

There are two quantities you'll likely be curious about. One is your acute change after an intervention, the other is your longer-term change—what I will call the "day after" effect, for the following example. The acute change is the difference between whatever your mental performance score was on the test you took right before your intervention and the score you got right after the intervention. If the second score was lower than the first, you might infer that the intervention depleted your abilities acutely. To see whether this pattern holds up across your whole experience, you should graph all of the acute differences from the whole experiment. To see longer-term changes, you'll want to take a different measure. When you take the same mental performance assessment at the beginning of your next neurohacking session, you'll now have a measure of the longer-term effect of what the intervention did—a "day after" effect. You can compare that number to the previous session's pre-intervention score. Then, just as you did for the acute differences, graph all of the "day after" numbers for your whole experiment. What patterns do you see? Let's look at an example. On Monday, your pre-intervention score was 10, your intervention was exercise, and your post-intervention score was 8. That would be an acute drop of 2. On Tuesday, your pre-intervention score was 12. Compared to the 10 you got for the pre-intervention score on Monday, your "day after" effect was an increase of 2.

Ultimately, you'll want to compare the performance of the two interventions against each other—assuming you used an a/b test design for your experiment. Essentially, the effects of each intervention are acting like the control group for the other intervention. To compare their performances against each other, you'll graph the "acute differences" and the "day after" effects of each intervention and see how they compare.

To know how much of an effect each intervention had on you, and whether that effect was due to anything more than chance, you'll need to use some statistical tools. Check out my website for additional recommendations to make analysis of your neurohacking experiments more intuitive. Below are examples of graphs you could end up with in an a/b type experiment. The first shows raw data from an alternating randomization design—it shows data collected during the intervention phase. The second graph shows the average acute effects of intervention A compared to intervention B.

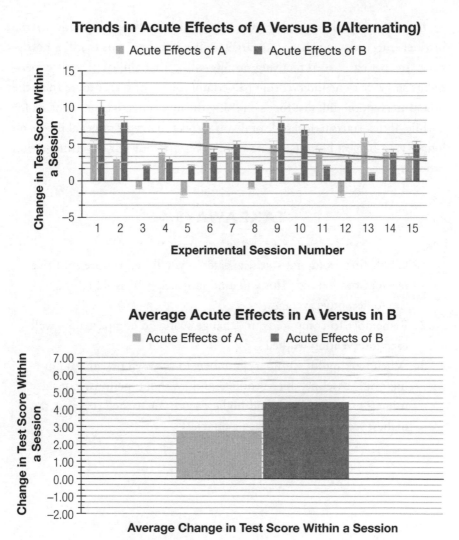

Trends in Acute Effects of A Versus B (Alternating)

■ Acute Effects of A   ■ Acute Effects of B

Change in Test Score Within a Session

Experimental Session Number

Average Acute Effects in A Versus in B

■ Acute Effects of A   ■ Acute Effects of B

Change in Test Score Within a Session

Average Change in Test Score Within a Session

Decisions and Next Steps

During your reflection, you'll have come to understand how well your self-experiment worked. You'll have discovered which interventions helped improve your mental target more immediately. Since you reassessed yourself across all the mental areas, across your health and lifestyle, and across your Life Satisfaction and Say to Do scores, you also have a sense of how much you changed more broadly.

Let's say you discover that you improved overall across many of the broader categories and that your target ability improved but is still a bottleneck. In that case, keep focusing on upgrading that ability. If your experiments show that one intervention helped improve your target more than the other intervention did, just focus on using the more effective intervention for a while. If that starts to feel stale or if you suspect that your target ability is no longer your biggest bottleneck, it's time to restart your self-experimentation process.

TAKEAWAYS

1. Graph first, compute statistics later—you'll get a more intuitive feel for how you did. This will especially help you avoid falling victim to deceptive averages.
2. Remember to compute your "change score" to compare how well each intervention worked.
3. When looking for summary numbers to convey how you did during your baseline, intervention, and washout periods, consider using averages or the second-highest score you earned (because the highest score could have been a fluke).

Final Goodies, Goodbye, and Good Luck!

"Science is organized knowledge. Wisdom is organized life."
—Immanuel Kant

> **Time investment:** 5 minutes
> **Goal:** To remind you of how far you've come and leave you with some final tools for the adventure ahead!

Congratulations, you've taken the first steps in your neurohacking journey! In this chapter we will recap the process we've discussed. You'll also learn about communities of other neurohackers to join. You'll be left with some final thoughts on how to be a good citizen neurohacker.

You've come a long way.

In Part I, you heard stories of neurohackers of all stripes. You learned that we have a surprising amount of control over the trajectory of our own brain development. You learned that there are significant differences between each of our brains and even the same brain over time. You learned how that means that self-experimentation is key to making sure that our brains change in ways we want them to. Ultimately, you learned how to run a self-experiment.

In Part II, you learned about four potential targets for a brain upgrade: executive function, emotional regulation, memory and learning, and creativity. You learned that not only can each of these areas be improved by teachers and doctors, but you can improve these abilities on your own, at home. You also learned that there are at-home tests you can use to gauge your progress over time on these abilities. In addition to these four mental targets, you learned about a host of potential causes for mental inefficiency that could be hidden away in your health or lifestyle; identifying such root causes could lead to a mental upgrade of its own. Finally, you learned about real-world outcomes you can track alongside your progress in your mental abilities. You learned how to track two examples of such real-world outcomes: your Life Satisfaction score and your Say to Do score.

In Part III, you learned about foundational interventions—the tools to experiment with before you try anything else. You looked at how to use the placebo effect to your advantage. You read about how blue light can work better than coffee, how recording your brain waves and learning how to manipulate them can significantly improve your mental performance. We looked at how playing games can make you smarter.

In Part IV, you learned about more advanced and, in some cases, riskier interventions. You learned how zapping your head with electricity might not be quite as foolish as it sounds, what kinds of products you can take to make your brain run more smoothly, and what technologies are already coming down the innovation pipeline.

In Part V, you found the 15-minute protocols for each of the interventions, organized by mental target, and guidance about how to use them.

Now that you're neurohacking, please visit my website to get more support on your neurohacking journey. You'll find additional tests and surveys, recommendations on interventions to try, and organizational tools to help you gather and make sense of your own mental performance data.

Some top tips:

1. Find a neurohacking buddy to hold you accountable and to make the process even more fun. To keep your exact cognitive data private, share just percent changes rather than your raw scores as you neurohack. Do learn and discuss your projects with friends and family, but remember not to make direct comparisons between your progress and your friends'. You're on your own path, and it will be an exciting one.

2. Connect with neurohacker communities around the world. If there's no group in your area, start one!

3. Please remember to stay safe, take care of yourself, and, if you involve others—which I highly encourage!—make sure to take care of them, too.

4. Remember that all brains are different, and they are changing all the time. There will be things that come easily to you and are hard for others and vice versa. There will be things that were hard for you last year that will be suddenly, inexplicably, easy for you this year.

5. Expect to be surprised. There's no need to limit yourself by other people's perceptions of you—and remember not to limit others by your perceptions of them. This is all very new, and there is much more that we don't know about our brains than that we do, so we need to all stay curious, bold, and kind throughout our journeys.

The brain is one of the most fascinating and ever-changing phenomena in the universe—and there is so much we have yet to learn about it. You can be part of that greater enterprise by exploring and becoming a master of your unique brain. Know that self-experimentation alternates between euphorically fulfilling and nail-bitingly frustrating, depending on the day.

I hope this book proves to be your friend and guide for years to come. Upgrading yourself is a process, and it never really has to end. Because there are so many interventions and mental targets to explore, I hope you reread these pages and try many different self-experiments. As neurofeedback pioneer Siegfried Othmer once said to me, "The failure to see gains with any particular method should not derail your search."[1] Brain change is a scientific reality, but it is not a given. It will be up to you to test different interventions until you find the right ones for you. You may find that the best intervention for you lies beyond these pages, perhaps with the help of a clinician, a teacher, or a friend. It may not be just one intervention, either; the "kitchen sink" approach of trying many interventions at once may be your best bet. Whatever approach you explore, I hope you subject it to the rigors and joys of self-experimentation.

Although neurohacking is a journey aimed at the self—greater self-discovery, self-understanding, and ultimately, self-mastery—you can choose to work on your own or share the journey with a buddy. And the most important thing? What you end up *doing* with that beautiful, newly upgraded brain of yours. Please stay in touch—tell me about your adventures.

You'll find many of the resources, videos, and exercises mentioned throughout the book on my website: ericker.com.

Happy neurohacking!

Elizabeth R. Ricker

P.S. If you enjoyed the book, please tell someone else about it and review it online so more people can join us in this movement!

Acknowledgments

"I not only use all the brains that I have, but all I can borrow."
—Woodrow Wilson

Any mistakes in this book are my own. Much of the good stuff came from conversations with the clever folks below.

Family

■ Varun: That such a dazzlingly inventive and deeply lovable human being believed in me and in this project helped me believe in it, too. For your patience and amusement at my repetitive questions about statistics and data analysis—and for the beautiful simulations that you computed for the self-experiment schedule stopping time. For forcing me to go to bed when I didn't want to. For making me laugh when I didn't want to. For watching our little munchkin so that I could write and edit, even when you had so many important pressures of your own—yet, somehow, still managing to excel at work. Thank you: I love you.

■ Mom and Dad: For instilling in me a love of learning. For wholeheartedly encouraging me in my projects, even while juggling so many important ones of your own. For supporting this project in every possible way. For those times when the rest of my earlier vision got lost in the trees of too many drafts and revisions, your love always brought me home.

■ Mom: For steeping me in your vision of the mind-body connection. For teaching me to seek a life structured around flow and an alignment of the heart, head, and hands. For teaching me how to pick myself up when I fell down—and for your legendary pep talks: hilarious, sometimes appalling, always motivating. For being the kind of trailblazing woman I hope to be when I (someday) grow up. For proving that being a badass lady and a woman with curiosity, kindness, wonder, and cheeky mischief are not mutually exclusive.

■ Dad: For stoking my love of the experimental method and always challenging me to figure things out for myself. For encouraging me to let this project marinate in the background for a while; it unstuck me and set this project on an evolutionary path that I couldn't have predicted. For being living proof that when big ideas plus a reverence for detail and an unstoppable work ethic meet, wild dreams can happen. For being the scientific conscience in my mind that seeks first principles, loathes sloppy thinking, and always comes back to the primary source. For teaching me how to fight insecurity by proving—sometimes mathematically—that our human perspective is often way, way too small.

■ Lindsey: For the many, many conversations over the years that challenged how I think about thinking. For being the big sister who taught me that the best way to enter cold water is with a cannonball. For inspiring me to write with more emotional honesty and to find my own voice. For the I-don't-even-know-how-many hours you spent reading this manuscript—and for the hilarious and right-on comments you sprinkled across every page.

■ Latha: Without your help with Munchkin, I could not have finished the manuscript. Your months of sacrifice meant that I could focus on my work knowing that he was safe and so adoringly cherished: thank you. As a side note, your cooking deserves a book of its own!

■ GP: Your questions and comments on the manuscript were not just encouraging, they were enriching. Your joyful devotion to our little guy is an inspiration to see.

■ Latha & GP: You are some of the most dedicated and generous people—in both spirit and action—that I've had the privilege of knowing. Thank you for our many free-ranging dinnertime discussions, for laugh-filled carrom games, for your loving care of our little one, and for your invaluable support as I finished up this project.

■ Munchkin: For bringing such unexpected joy and meaning to every day since your birth. Also, for helping me write this book before you even took your first breath—how many babies can claim that!? For all of your smiles, coos, and mischief. I can't wait to run experiments (on you? with you?) when you're older...

Team

■ Howard Yoon: For sticking with this project with me through all the years. And for fighting for my book at every turn, even through injury, Covid, birth and death, and everything else the world threw at us.

■ Marisa Vigilante: For understanding what I was trying to do from the very beginning, for terrific structural insights on the manuscript, and for being understanding even as the pandemic and (both of our) pregnancies threw wrenches in the schedule.

■ To Tracy Behar: Thank you for this opportunity—and for taking a gamble on me as an author. To Ian Straus, for countless patient and detailed responses to my many, many questions. To the entire Little, Brown Spark/ HBG team: thank you for turning my pipe dream into a reality.

■ Hannah Kushnick: For being a deep and broad thinker, an insightful editor, and a big-hearted confidante; working with you was one of my best career decisions. You went the extra ten miles, you made me laugh instead of pulling my hair out, and you somehow sensed what I was *really* trying to say—and helped tease it onto the page.

■ Stephanie Fine Sasse: For being the best accountability buddy I've ever had. For so instantly and deeply understanding what I was trying to do with this book, for great feedback on the manuscript, for loving spreadsheets and self-tracking just as much as I do—and for drawing some badass brain pics.

■ The Science Fact Checking and Citations Team: Thank you so much for your brilliance, hard work, and loving attention to the science that underpins all of these adventures. What a relief to know this phenomenal team was double-checking my work! You caught ambiguities, hunted down research papers, and asked terrific questions: Samantha Bureau, Manoj Doss, Ritu

Gaikwad, Azure Grant, Gabriella (Ella) Hirsch, Katherine Miclau, Noah Grey Rosenzweig, Alea Skwara, Laetitia Wang.

■ My Neurdy "Brain Trust" (see what I did there?)—Irina Skylar-Scott, Chung-Hay Luk, Neville Sanjana, Amy Daitch: What a privilege to have such a crew reading my manuscript: thank you for your sharp eyes, the additional outside research you introduced me to, and for your enthusiasm for this project.

■ All of the subjects, all of the innovators, and all of the researchers whose work fills the pages of this book: I hope I've done justice to your work. I also hope that I've filled the reader with the same delight and respect I hold for you. Thank you for what you do. Thank you for letting me share some of your stories and findings with the world.

Mentors/inspirations/friends

■ Todd Rose: For being the most tolerant, understanding, and genuinely supportive grad school adviser a kid could ask for.

■ Ogi Ogas and Tina Seelig: For believing in this book from the very beginning and for providing encouragement, feedback, and support at pivotal moments.

■ Anna Weiss: For being a phenomenal sounding board—enthused, broad in perspective, and warm. And for texting me partway through reading the first draft to tell me you loved it; I think that text gave me years of my life back.

■ David Eagleman (and Lindsey for introducing us): What luck to have met such an intellectual live wire (sorry, I couldn't resist that one) so early in my life! Thank you for being the best kind of polymath role model with your awe-inspiring energy and your generous advice and encouragement, and for your continued interest in this project throughout the years.

■ Camilla Rockefeller: For your everlasting curiosity and delightfully imaginative questions about all things mind and brain related; imagining how I'd explain a finding to you helped me explain some of the trickier ideas in this book. For rewarding so many of my bad jokes with gales of laughter.

For all-night sleepovers that zigzagged between the philosophical and the absurd, starting at age nine—and pretty much not stopping ever. For your truly helpful comments on the first draft too.

■ Pia Pal: For so many fascinating conversations about brains, organization, and productivity.

■ David Snider: For conversations that challenged me to believe anything was possible, even when we were just middle and high schoolers. Right after you'd finished publishing your book, you told me you looked forward to reading mine—I never forgot that.

■ So many others who played roles large and small in making this creature come to life.

■ Cousin Rozzie/Rosalind Berlin: I wish you had lived long enough for me to discuss this project with you. Your scalpel-sharp questions would have sent me back to the primary sources dozens of times...but it would have been so much fun.

■ Sugar Magnolia Peeps: Rob—for keeping a roof over my head even as San Francisco's rents skyrocketed; Emily—for proving to me that gluten intolerance doesn't have to mean the end of delicious food; Altay—for amazing music and delightful philosophical meanderings; to my other housemates: for being kind, funny, creative, and for throwing epic parties (here's looking at you, Courtney).

Notes

Author's Note

1. The convention I use is to convert a Cohen's D effect size into a percentile standing. Technically, when I say "percent," it is really "percentile point on average." For more, go to "Effect Size (ES) | Effect Size Calculators." n.d. lbecker.uccs.edu/effect-size.

Introduction

1. Name changed.
2. Shaywitz, S. E., et al. "Persistence of Dyslexia: The Connecticut Longitudinal Study at Adolescence." *Pediatrics* 104 (6): 1351–59, 1999. doi:10.1542/peds.104.6.1351.
3. Fiester, Leila, and Ralph Smith. "Early Warning! Why Reading by the End of Third Grade Matters." A KIDS COUNT Special Report from the Annie E. Casey Foundation. 2010. files.cric.ed .gov/fulltext/ED509795.pdf; Planty, Michael, et al. "The Condition of Education 2009. NCES 2009-081." National Center for Education Statistics: 41, 2009.
4. Saygin, Z. M., et al. "Tracking the Roots of Reading Ability: White Matter Volume and Integrity Correlate with Phonological Awareness in Prereading and Early-Reading Kindergarten Children." *Journal of Neuroscience* 33 (33): 13251–58, 2013. doi:10.1523/jneurosci.4383-12.2013; Centanni, Tracy M., et al. "Disrupted Left Fusiform Response to Print in Beginning Kindergartners Is Associated with Subsequent Reading." *NeuroImage: Clinical* 22: 101715, 2019. doi:10.1016/j.nicl.2019.101715.
5. Rabiner, David, and John D. Coie. "Early Attention Problems and Children's Reading Achievement: A Longitudinal Investigation." *Journal of the American Academy of Child and Adolescent Psychiatry* 39 (7): 859–67, 2000. ncbi.nlm.nih.gov/pmc/articles/PMC2777533/.

Chapter 1: Scientific Self-Help

1. Research and Markets. "The U.S. Market for Self-Improvement Products & Services, 2003–2023: Market Size & Growth, Trends, In-Depth Profiles of 60 Top Motivational Speakers, List of the Top 100 Infomercials." GlobeNewswire News Room. March 3, 2020. globenewswire .com/news-release/2020/03/03/1994097/0/en/The-U-S-Market-for-Self-improvement -Products-Services-2003-2023-Market-Size-Growth-Trends-In-Depth-Profiles-of-60-Top -Motivational-Speakers-List-of-the-Top-100-Infomercials.html.
2. Hanley, Brian P., William Bains, and George Church. "Review of Scientific Self-Experimentation: Ethics History, Regulation, Scenarios, and Views Among Ethics Committees and Prominent Scientists." *Rejuvenation Research* 22 (1): 31–42, 2019. doi:10.1089/rej.2018.2059.
3. Heidt, Amanda. "Self-Experimentation in the Time of COVID-19." *The Scientist Magazine.* the -scientist.com/news-opinion/self-experimentation-in-the-time-of-covid-19-67805; Tan, Sy, and

N. Ponstein. "Jonas Salk (1914–1995): A Vaccine against Polio." *Singapore Medical Journal* 60 (1): 9–10, 2019. doi:10.11622/smedj.2019002.

4. Altman, Lawrence K. *Who Goes First?: The Story of Self- Experimentation in Medicine.* Berkeley: University of California Press, 1998.

Chapter 2: Neurohackers, Revealed

1. "Roger Craig—Spaced Repetition: A Cognitive QS Method for Knowledge Acquisition." Vimeo. August 23, 2012. vimeo.com/48070812.

2. Wolf, Gary. "Want to Remember Everything You'll Ever Learn? Surrender to This Algorithm." *Wired.* April 21, 2008. wired.com/2008/04/ff-wozniak/.

3. For more on one of the key pioneers of modern spaced repetition, Piotr Wozniak, and his SuperMemo software: wired.com/2008/04/ff-wozniak/.

4. Craig, Roger. n.d. "Knowledge Tracking." Quantified Self. quantifiedself.com/show-and-tell /?project=638.

5. Harrison, Richard. n.d. "How I Lost 200 Pounds." Quantified Self. quantifiedself.com/show -and-tell/?project=607.

6. Boesel, Whitney Erin. n.d. "My Numbers Sucked but I Made This Baby Anyway." Quantified Self. quantifiedself.com/show-and-tell/?project=1079.

7. Jonas, Steven. n.d. "Show & Tell Projects Archive." Quantified Self. quantifiedself.com/show -and-tell/?project=213.

8. Drangsholt, Mark. "Deciphering My Brain Fog." Quantified Self. August 21, 2014. quantifiedself .com/blog/mark-drangsholt-deciphering-brain-fog/.

Chapter 3: The Evidence

1. Kaufman, Alan S. *IQ Testing 101.* New York, London: Springer, 2009.

2. Kaufman. *IQ Testing 101.*

3. Kim, Woojong, et al. "An FMRI Study of Differences in Brain Activity among Elite, Expert, and Novice Archers at the Moment of Optimal Aiming." *Cognitive and Behavioral Neurology* 27 (4): 173–82, 2014. doi.org/10.1097/wnn.0000000000000042.

4. Draganski, B., et al. "Temporal and Spatial Dynamics of Brain Structure Changes during Extensive Learning." *Journal of Neuroscience* 26 (23): 6314–17, 2006. doi.org/10.1523/jneurosci .4628-05.2006.

5. Zatorre, Robert J., R. Douglas Fields, and Heidi Johansen-Berg. "Plasticity in Gray and White: Neuroimaging Changes in Brain Structure during Learning." *Nature Neuroscience* 15 (4): 528–36, 2012. doi.org/10.1038/nn.3045.

6. Woollett, Katherine, and Eleanor A. Maguire. "Acquiring 'the Knowledge' of London's Layout Drives Structural Brain Changes." *Current Biology* 21 (24): 2109–14, 2011. doi.org/10.1016/j .cub.2011.11.018.

7. Mackey, Allyson P., Kirstie J. Whitaker, and Silvia A. Bunge. "Experience-Dependent Plasticity in White Matter Microstructure: Reasoning Training Alters Structural Connectivity." *Frontiers in Neuroanatomy* 6 (32), 2012. doi.org/10.3389/fnana.2012.00032.

8. Draganski, et al. "Temporal and Spatial Dynamics."

9. Seminowicz, David A., et al. "Cognitive-Behavioral Therapy Increases Prefrontal Cortex Gray Matter in Patients with Chronic Pain." *Journal of Pain* 14 (12): 1573–84, 2013. doi.org/10.1016/j .jpain.2013.07.020.

10. Yeh, Fang-Cheng, et al. "Quantifying Differences and Similarities in Whole-Brain White

Matter Architecture Using Local Connectome Fingerprints." *PLOS Computational Biology* 12 (11): e1005203, 2016. doi.org/10.1371/journal.pcbi.1005203.

11. Paul Pringle. "College Board Scores with Critics of SAT Analogies." *Los Angeles Times.* July 27, 2003. latimes.com/archives/la-xpm-2003-jul-27-me-sat27-story.html.

12. "De Novo Classification Request for Neuropsychiatric EEG-Based Assessment Aid for ADHD (NEBA) System Regulatory Information." n.d. accessdata.fda.gov/cdrh_docs/reviews/K112711.pdf.

13. Valizadeh, Seyed Abolfazl, et al. "Identification of Individual Subjects on the Basis of Their Brain Anatomical Features." *Scientific Reports* 8 (1), 2018. doi.org/10.1038/s41598-018-23696-6.

14. Mars, Rogier B., Richard E. Passingham, and Saad Jbabdi. "Connectivity Fingerprints: From Areal Descriptions to Abstract Spaces." *Trends in Cognitive Sciences* 22 (11): 1026–37, 2018. doi .org/10.1016/j.tics.2018.08.009.

15. Yeh, et al. "Quantifying Differences."

16. Ruiz-Blondet, Maria V., Zhanpeng Jin, and Sarah Laszlo. "CEREBRE: A Novel Method for Very High Accuracy Event-Related Potential Biometric Identification." *IEEE Transactions on Information Forensics and Security* 11 (7): 1618–29, 2016. doi.org/10.1109/TIFS.2016.2543524.

17. Armstrong, Thomas. *The Power of Neurodiversity: Discovering the Extraordinary Gifts of Autism, ADHD, Dyslexia, and Other Brain Differences.* Cambridge, MA: Da Capo Lifelong, 2010.

18. Whipps, Heather. "Why Did People Become White?" Livescience.com. September 1, 2009. livescience.com/7863-people-white.html.

19. Posner, David. "ADHD: An Overview by David Posner, M.D.—ADD Resource Center." September 23, 2012. addrc.org/adhd-an-overview/#:~:text=ADHD%20is%20more%20common%20in.

20. Facer-Childs, Elise R., Sophie Boiling, and George M. Balanos. "The Effects of Time of Day and Chronotype on Cognitive and Physical Performance in Healthy Volunteers." *Sports Medicine Open* 4 (1), 2018. doi.org/10.1186/s40798-018-0162-z.

Chapter 4: The Nuts and Bolts

1. Othmer, Siegfried. Letter to Elizabeth Ricker. Email, March 15, 2021.

2. "Quantified Mind Team." n.d. quantified-mind.com/team.

3. Donner, Yoni. n.d. "Quantified Mind: Scalable Assessment of Within-Person Variation in Cognition." forum.stanford.edu/events/posterslides/QuantifiedMindEfficientScalableAssessment ofWithinpersonVariationinCognitiveAbilities.pdf.

4. Ngandu, Tiia, et al. "A 2 Year Multidomain Intervention of Diet, Exercise, Cognitive Training, and Vascular Risk Monitoring Versus Control to Prevent Cognitive Decline in At-Risk Elderly People (FINGER): A Randomised Controlled Trial." *Lancet* 385 (9984): 2255–63, 2015. doi .org/10.1016/S0140-6736(15)60461-5.

5. "Neuroscience Lightning Talks." n.d. youtube.com/watch?v=DzwvaicdOJU.

6. "Introduction to Cross-over Designs." n.d. Penn State: Statistics Online Courses. online.stat .psu.edu/stat502/lesson/11/11.1.

Chapter 5: Organize to Motivate

1. "Goals Research Summary." n.d. dominican.edu/sites/default/files/2020-02/gailmatthews -harvard-goals-researchsummary.pdf.

2. Nickerson, David W., and Todd Rogers. "Do You Have a Voting Plan?: Implementation Intentions, Voter Turnout, and Organic Plan Making." *Psychological Science* 21 (2): 194–99, 2010. journals.sagepub.com/doi/abs/10.1177/0956797609359326.

3. Nolan, Christopher. *Batman Begins.* Warner Bros. Pictures, DC Comics, Legendary Pictures, Syncopy, Patalex III Productions: 2006.

4. Milkman, Katherine L. "The Science of Keeping Your New Year's Resolution." *Washington Post,* January 1, 2018. washingtonpost.com/news/wonk/wp/2018/01/01/the-science-of-keeping-your -new-years-resolution/.

5. Sharif, Marissa A., and Suzanne B. Shu. "The Benefits of Emergency Reserves: Greater Prefer- ence and Persistence for Goals That Have Slack with a Cost." *Journal of Marketing Research* 54 (3): 495–509, 2017. doi.org/10.1509/jmr.15.0231.

Chapter 6: Debugging Yourself

1. Baldacci, S., et al. "Allergy and Asthma: Effects of the Exposure to Particulate Matter and Bio- logical Allergens." *Respiratory Medicine* 109 (9): 1089–104, 2015. doi:10.1016/j.rmed.2015.05.017.

2. Killgore, William, D. S. "Effects of Sleep Deprivation on Cognition." *Progress in Brain Research* 185: 105–29, 2010. doi:10.1016/b978-0-444-53702-7.00007-5.

3. Shehab, M. A., and F. D. Pope. "Effects of Short-Term Exposure to Particulate Matter Air Pollu- tion on Cognitive Performance." *Scientific Reports* 9 (1), 2019. doi:10.1038/s41598-019-44561-0.

4. Ingraham, Christopher. "Heat Makes You Dumb, in Four Charts." *Washington Post.* July 7, 2018. washingtonpost.com/business/2018/07/17/heat-makes-you-dumb-four-charts; Goodman, Joshua, et al. "Heat and Learning." National Bureau of Economic Research. 2018. nber.org /papers/w24639.

5. Selhub, Eva. "Nutritional Psychiatry: Your Brain on Food." Harvard Health Blog. April 5, 2018. health.harvard.edu/blog/nutritional-psychiatry-your-brain-on-food-201511168626.

6. Spencer, Sarah J., et al. "Food for Thought: How Nutrition Impacts Cognition and Emotion." *NPJ Science of Food* 1 (1), 2017. doi:10.1038/s41538-017-0008-y; Vojdani, Aristo, Datis Kharrazian, and Partha Mukherjee. "The Prevalence of Antibodies against Wheat and Milk Proteins in Blood Donors and Their Contribution to Neuroimmune Reactivities." *Nutrients* 6 (1): 15–36, 2013. doi:10.3390/ nu6010015; Yelland, Gregory W. "Gluten-Induced Cognitive Impairment ('Brain Fog') in Coeliac Disease." *Journal of Gastroenterology and Hepatology* 32: 90–93, 2017. doi:10.1111/jgh.13706.

7. Spencer, et al. "Food for Thought."

8. Peters, Steven. "Nearly 10% of Americans Have a Nutritional Deficiency. These Are the Most Common." *USA Today.* August 20, 2019. usatoday.com/picture-gallery/news/health/2019 /08/20/most-common-nutritional-deficiencies-iron-copper-calcium-among-top/39976199/.

9. "CDC'S Second Nutrition Report: A Comprehensive Biochemical Assessment of the Nutrition Status of the U.S. Population." CDC. 2012. cdc.gov/nutritionreport/pdf/Nutrition_Book _complete508_final.pdf.

10. Grinder-Pedersen, et al. "Calcium from Milk or Calcium-Fortified Foods Does Not Inhibit Nonheme-Iron Absorption from a Whole Diet Consumed over a 4-d Period." *American Journal of Clinical Nutrition* 80 (2): 404–9, 2004. doi:10.1093/ajcn/80.2.404.

11. Rae, Caroline, et al. "Oral Creatine Monohydrate Supplementation Improves Brain Perfor- mance: a Double-Blind, Placebo-Controlled, Cross-over Trial." *Proceedings of the Royal Society of London. Series B: Biological Sciences* 270 (1529): 2147–50, 2003. doi:10.1098/rspb.2003.2492.

12. Rae, et al. "Oral Creatine Monohydrate."

13. Brown, Wyatt. "Do You Need to Take Multivitamins?" Examine.com. September 18, 2019. examine.com/nutrition/do-you-need-a-multivitamin/; Calton, Jayson B. "Prevalence of Micro- nutrient Deficiency in Popular Diet Plans." *Journal of the International Society of Sports Nutrition* 7 (1), 2010. doi.org/10.1186/1550-2783-7-24; Vici, Giorgia, et al. "Gluten Free Diet and Nutri- ent Deficiencies: A Review." *Clinical Nutrition* 35 (6): 1236–41, 2016. doi.org/10.1016/j .clnu.2016.05.002; Craig, Winston J. "Health Effects of Vegan Diets." *American Journal of Clin- ical Nutrition* 89 (5): 1627S1633S, 2009. htdoi.org/10.3945/ajcn.2009.26736n.

14. Harvard Health Publishing. "Should You Get Your Nutrients from Food or from Supplements?" Harvard Health. 2015. health.harvard.edu/staying-healthy/should-you-get-your-nutrients-from -food-or-from-supplements.

15. Clifford, J., and Curely, J. "Water-Soluble Vitamins: B-Complex and Vitamin C-9.312." *Extension.* February 26, 2020. extension.colostate.edu/topic-areas/nutrition-food-safety-health /water-soluble-vitamins-b-complex-and-vitamin-c-9-312/; Weil, Andrew. "Can Any B Vitamins Be Harmful?" DrWeil.com. 2019. drweil.com/vitamins-supplements-herbs/vitamins/can-any -b-vitamins-be-harmful/.

16. Forbes, Scott C., et al. "Effect of Nutrients, Dietary Supplements and Vitamins on Cognition: A Systematic Review and Meta-Analysis of Randomized Controlled Trials." *Canadian Geriatrics Journal* 18 (4), 2015. doi.org/10.5770/cgj.18.189.

17. Mazereeuw, Graham, et al. "Effects of Omega-3 Fatty Acids on Cognitive Performance: a Meta-Analysis." *Neurobiology of Aging* 33 (7), 2012. doi:10.1016/j.neurobiolaging.2011.12.014.

18. Callahan, Alice. 2017. "Do DHA Supplements Make Babies Smarter?" *New York Times.* nytimes.com/2017/03/30/well/do-dha-supplements-make-babies-smarter.html.

19. Hoover, Nathan, Ashlyn Aguiniga, and Jaime Hornecker. "In the Adult Population, Does Daily Multivitamin Intake Reduce the Risk of Mortality Compared with Those Who Do Not Take Daily Multivitamins?" *Evidence-Based Practice* 22 (3): 15, 2019. doi:10.1097/ebp.0000000000000186.

20. Mursu, Jaakko, et al. "Dietary Supplements and Mortality Rate in Older Women." *Archives of Internal Medicine* 171 (18): 1625, 2011. doi:10.1001/archinternmed.2011.445.

21. Jairoun, Ammar Abdulrahman, Moyad Shahwan, and Sa'Ed H. Zyoud. "Heavy Metal Contamination of Dietary Supplements Products Available in the UAE Markets and the Associated Risk." *Scientific Reports* 10 (1), 2020. doi:10.1038/s41598-020-76000-w.

22. Buscemi, Nina, et al. "Melatonin for Treatment of Sleep Disorders: Summary." *AHRQ Evidence Report Summaries.* 2004.

23. Skylar-Scott, Irina. Letter to Elizabeth Ricker. February 14, 2021; Pressler, Ann. "Melatonin: How Much Should I Take for a Good Night's Rest?" *Health Essentials from Cleveland Clinic.* 2020. health.clevelandclinic.org/melatonin-how-much-should-i-take-for-a-good-nights-rest/.

24. "Melatonin: What You Need to Know." National Center for Complementary and Integrative Health. 2021. nccih.nih.gov/health/melatonin-what-you-need-to-know.

25. Harvard Health Publishing. "By the Way, Doctor: Are Sleeping Pills Addictive?" Harvard Health, n.d. health.harvard.edu/newsletter_article/By_the_way_doctor_Are_sleeping_pills_addictive.

26. Walker, Matthew P. *Why We Sleep: Unlocking the Power of Sleep and Dreams.* New York: Scribner, 2018; Fitzgerald, Timothy, and Jeffrey Vietri. "Residual Effects of Sleep Medications Are Commonly Reported and Associated with Impaired Patient-Reported Outcomes among Insomnia Patients in the United States." *Sleep Disorders* 2015: 1–9, 2015. doi:10.1155/2015/607148.

27. Willyard, Cassandra. "How Gut Microbes Could Drive Brain Disorders." *Nature* 590 (7844): 22–25, 2021. doi.org/10.1038/d41586-021-00260-3.

28. Quintero, Esther. "Feeling Socially Connected Fuels Intrinsic Motivation and Engagement." Shanker Institute. 2015. shankerinstitute.org/blog/feeling-socially-connected-fuels-intrinsic -motivation-and-engagement.

29. Beutel, Manfred E., et al. "Loneliness in the General Population: Prevalence, Determinants and Relations to Mental Health." *BMC Psychiatry* 17 (1), 2017. doi:10.1186/s12888-017-1262-x.

30. "Facts and Statistics." Anxiety and Depression Association of America (ADAA). 2021. adaa .org/understanding-anxiety/facts-statistics.

31. Maletic, V., et al. "Neurobiology of Depression: An Integrated View of Key Findings." *International Journal of Clinical Practice* 61 (12): 2030–40, 2007. doi:10.1111/j.1742-1241.2007.01602.x.

32. Sackeim, Harold A., et al. "Effects of Major Depression on Estimates of Intelligence." *Journal of Clinical and Experimental Neuropsychology* 14 (2): 268–88, 1992. doi:10.1080/01688639208402828; Payne, Tabitha W., and Madeline Thompson. "Impaired Mental Processing Speed with Moderate to Severe Symptoms of Depression." *Major Depressive Disorder: Cognitive and Neurobiological Mechanisms.* 2015. doi:10.5772/59597.

33. Mandelli, Laura, et al. "Improvement of Cognitive Functioning in Mood Disorder Patients with Depressive Symptomatic Recovery during Treatment: An Exploratory Analysis." *Psychiatry and Clinical Neurosciences* 60 (5): 598–604, 2006. doi:10.1111/j.1440-1819.2006.01564.x; Priyamvada, Richa, Rupesh Ranjan, and Suprakash Chaudhury. "Cognitive Rehabilitation of Attention and Memory in Depression." *Industrial Psychiatry Journal* 24 (1): 48, 2015. doi:10.4103/0972-6748.160932; Schaefer, Jonathan D., et al. "Is Low Cognitive Functioning a Predictor or Consequence of Major Depressive Disorder? A Test in Two Longitudinal Birth Cohorts." *Development and Psychopathology:* 1–15, 2017. doi:10.1017/s095457941700164x.

34. "Mental Illness." National Institute of Mental Health. U.S. Department of Health and Human Services. 2021.nimh.nih.gov/health/statistics/mental-illness.shtml.

35. "Mental Health By the Numbers." National Alliance on Mental Illness (NAMI). 2012. nami.org/mhstats.

36. "The 9 Best Online Therapy Programs of 2019." Verywell Mind. 2019. verywellmind.com/best-online-therapy-4691206.

37. Puhan, Milo A, et al. "Didgeridoo Playing as Alternative Treatment for Obstructive Sleep Apnoea Syndrome: Randomised Controlled Trial." *British Medical Journal* 332 (7536): 266–70, 2006. doi.org/10.1136/bmj.38705.470590.55.

38. Lahl, Olaf, et al. "An Ultra Short Episode of Sleep Is Sufficient to Promote Declarative Memory Performance." *Journal of Sleep Research* 17 (1): 3–10, 2008. onlinelibrary.wiley.com/doi/full/10.1111/j.1365-2869.2008.00622.x.

39. "Fall Asleep in 120 Seconds." healthline.com/health/healthy-sleep/fall-asleep-fast#foundation-alsleep-tips and washingtonpost.com/lifestyle/2019/01/08/i-stopped-sleeping-then-i-tried-method-thats-supposed-work-two-minutes-or-less/.

40. Carefoot, Helen. "Plastic, Metal or Glass: What's the Best Material for a Reusable Water Bottle?" *Washington Post.* 2019. washingtonpost.com/lifestyle/plastic-metal-or-glass-whats-the-best-material-for-a-reusable-water-bottle/2019/09/25/5edcbe6c-d957-11e9-bfb1-849887369476_story.html.

41. "9 out of 10 People Worldwide Breathe Polluted Air, But More Countries Are Taking Action." World Health Organization. 2018. who.int/news-room/detail/02-05-2018-9-out-of-10-people-worldwide-breathe-polluted-air-but-more-countries-are-taking-action.

42. qz.com/470301/men-are-literally-freezing-women-out-of-the-workplace/; Energy consumption in buildings and female thermal demand: nature.com/articles/nclimate2741; Metabolic equivalent: one size does not fit all: ncbi.nlm.nih.gov/pubmed/15831804 assumed a 154-pound, 40-yr old man (35% higher metabolism than women); comparison between male and female subjective estimates of thermal effects and sensations: ncbi.nlm.nih.gov/pubmed/15676395 71.6 degrees F (22 degrees C) vs 77.1 degrees F (25 degrees C) found in that source; arstechnica.com/science/2019/05/test-performance-gender-and-temperature/; hsph.harvard.edu/news/press-releases/extreme-heat-linked-with-reduced-cognitive-performance-among-young-adults-in-non-air-conditioned-buildings/; PSAT study: 1% lost for every 1 degree F increase (much higher effect in non-A/C areas, much higher effect among minority students): hks.harvard.edu/announcements/when-heat-student-learning-suffers, original study is here: scholar.harvard.edu/files/joshuagoodman/files

/w24639.pdf; Gaokao study in China: conference.iza.org/conference_files/environ_2019/graff %20zivin_j9945.pdf; NYC public schools and high school dropouts: scholar.harvard.edu/files /jisungpark/files/paper_nyc_aejep.pdf; Ingraham, Christopher. "Heat Makes You Dumb, in Four Charts." *Washington Post.* July 17, 2018. washingtonpost.com/business/2018/07/17/heat-makes -you-dumb-four-charts/.

43. Ingraham. "Heat Makes You Dumb."

44. Marshall, P. S., and E. A. Colon. "Effects of Allergy Season on Mood and Cognitive Function." *Annals of Allergy, Asthma & Immunology* 71 (3): 251–8, 1993. PMID: 8372999. pubmed.ncbi .nlm.nih.gov/8372999/.

45. An app called SpotCrime uses data from the police to map areas where crimes have occurred, alerting users so they can potentially avoid trouble spots: spotcrime.com/.

46. CDC. "Loneliness and Social Isolation Linked to Serious Health Conditions." May 26, 2020. cdc.gov/aging/publications/features/lonely-older-adults.html.

47. Killam, Kasley. "To Combat Loneliness, Promote Social Health." *Scientific American.* 2018. scientificamerican.com/article/to-combat-loneliness-promote-social-health1/.

48. Killam, Kasley. "A Solution for Loneliness." *Scientific American.* 2019. scientificamerican.com /article/a-solution-for-loneliness/.

49. O'Connell, Brenda H., Deirdre Oshea, and Stephen Gallagher. "Mediating Effects of Loneliness on the Gratitude-Health Link." *Personality and Individual Differences* 98: 179–83, 2016. doi:10.1016/j.paid.2016.04.042.

50. Käll, Anton, et al. "Internet-Based Cognitive Behavior Therapy for Loneliness: A Pilot Randomized Controlled Trial." *Behavior Therapy* 51 (1): 54–68, 2020. doi:10.1016/j.beth.2019.05.001.

51. Nortje, Alicia. "Realizing Your Meaning: 5 Ways to Live a Meaningful Life." PositivePsychology .com. 2021. positivepsychology.com/live-meaningful-life/; Smith, Jeremy Adam. "How to Find Your Purpose in Life." Greater Good. 2018. greatergood.berkeley.edu/article/item/how_to _find_your_purpose_in_life.

Chapter 7: The New IQ

1. Note that this chapter uses the term "new IQ" to describe executive function. Note that there is also a book by Tracy Alloway and Ross Alloway called *The New IQ: Use Your Working Memory to Think Stronger, Smarter, Faster* that is about working memory rather than executive function.

2. Diamond, D. M., et al. "The Temporal Dynamics Model of Emotional Memory Processing: A Synthesis on the Neurobiological Basis of Stress-Induced Amnesia, Flashbulb and Traumatic Memories, and the Yerkes-Dodson Law." *Neural Plasticity* 33, 2007. doi:10.1155/2007/60803. PMID 17641736. commons.wikimedia.org/w/index.php?curid=34030384.

3. Diamond, et al., "The Temporal Dynamics."

4. In addition to using MRI (which is a correlational tool), researchers have used causal methods to confirm the relationship between executive function and specific parts of the brain—causal methods such as TMS (transcranial magnetic stimulation).

5. Alvarez, Julie A., and Eugene Emory. "Executive Function and the Frontal Lobes: A Meta-Analytic Review." *Neuropsychology Review* 16 (1): 17–42, 2006. doi.org/10.1007/s11065-006-9002-x.

6. Fine Sasse, Stephanie. *Executive Function: Frontal Cortex.* 2021.

7. Moffitt, Terrie E., et al. "A Gradient of Childhood Self-Control Predicts Health, Wealth, and Public Safety." *Proceedings of the National Academy of Sciences* 108 (7): 2693–98, 2011.

8. Executive functions predict academic performance in the earliest elementary grades through university better than does IQ. (Alloway and Alloway, 2010; Bull and Scerif, 2001; Dumontheil

and Klingberg, 2012; Gathercole, et al., 2004; McClelland and Cameron, 2011; Nicholson, 2007; Passolunghi, et al., 2007; St Clair-Thompson and Gathercole, 2006; Savage, et al., 2006; Swanson, 2014). From slide 17 on pdx.edu/social-determinants-health/sites/www.pdx.edu .social-determinants-health/files/Adele%20Diamond.pdf.

9. "Christine Hooker." rushu.rush.edu/faculty/christine-hooker-phd.

10. Alloway, Tracy Packiam, and Ross G Alloway. "Investigating the Predictive Roles of Working Memory and IQ in Academic Attainment." *Journal of Experimental Child Psychology* 106: 20–29, 2010. doi.org/10.1016/j.jecp.2009.11.003.

11. Kaufman, Scott Barry. "The Mind of the Prodigy." *Scientific American.* February 10, 2014. blogs.scientificamerican.com/beautiful-minds/the-mind-of-the-prodigy/.

12. Germine, Laura, et al. "Is the Web as Good as the Lab? Comparable Performance from Web and Lab in Cognitive/Perceptual Experiments." *Psychonomic Bulletin & Review* 19 (5): 847–57, 2012. doi.org/10.3758/s13423-012-0296-9.

13. Feenstra, Heleen E. M., et al. "Online Self-Administered Cognitive Testing Using the Amsterdam Cognition Scan: Establishing Psychometric Properties and Normative Data." *Journal of Medical Internet Research* 20 (5): e192, 2018. doi.org/10.2196/jmir.9298.

14. Some inspiration from the following sources: Castellanos, Irina, William G. Kronenberger, and David B. Pisoni. "Questionnaire-Based Assessment of Executive Functioning: Psychometrics." *Applied Neuropsychology: Child* 7 (2): 93–109, 2016. doi.org/10.1080/21622965.2016.1248557; nyspta.org/wp-content/uploads/2017/08/Conv17-305-dawson-executive-skills-questionnaire .pdf; executivefunctionmatters.com/wp-content/uploads/2012/04/Executive-Function-Processes -Self-Assessment.pd.

15. Baddeley, Alan. "Working Memory: Looking Back and Looking Forward." *Nature Reviews Neuroscience* 4 (10): 829–39, 2003. doi.org/10.1038/nrn1201.

16. Del Rossi, Gianluca, Alfonso Malaguti, and Samanta Del Rossi. "Practice Effects Associated with Repeated Assessment of a Clinical Test of Reaction Time." *Journal of Athletic Training* 49 (3): 356–59, 2014. doi.org/10.4085/1062-6059-49.2.04.

17. Schooten, Kimberley S. van, et al. "Catch the Ruler: Concurrent Validity and Test–Retest Reliability of the ReacStick Measures of Reaction Time and Inhibitory Executive Function in Older People." *Aging Clinical and Experimental Research* 31 (8): 1147–54, 2018. doi.org/10.1007/s40520 -018-1050-6; Kail, Robert, and Timothy A. Salthouse. "Processing Speed as a Mental Capacity." *Acta Psychologica* 86 (2–3): 199–225, 1994. doi.org/10.1016/0001-6918(94)90003-5.

18. Eriksen, B. A., and C. W. Eriksen. "Effects of Noise Letters upon Identification of a Target Letter in a Non-Search Task." *Perception and Psychophysics* 16: 143–49, 1974. doi:10.3758/ bf03203267.

19. Stroop, J. R. "Studies of Interference in Serial Verbal Reactions." *Journal of Experimental Psychology* 18 (6): 643–62, 1935. doi.org/10.1037/h0054651.

20. Stroop, "Studies of Interference in Serial Verbal Reactions."

Chapter 8: The New EQ

1. Kelley, Nicholas J., et al. "Stimulating Self-Regulation: A Review of Non-Invasive Brain Stimulation Studies of Goal-Directed Behavior." *Frontiers in Behavioral Neuroscience* 12 (337), 2019. doi.org/10.3389/fnbeh.2018.00337.

2. Aldao, Amelia, Gal Sheppes, and James J. Gross. "Emotion Regulation Flexibility." *Cognitive Therapy and Research* 39 (3): 263–78, 2015. doi.org/10.1007/s10608-014-9662-4.

3. Urquijo, Itziar, Natalio Extremera, and Josu Solabarrieta. "Connecting Emotion Regulation to

Career Outcomes: Do Proactivity and Job Search Self-Efficacy Mediate This Link?" *Psychology Research and Behavior Management* 12 (December): 1109–20, 2019. doi.org/10.2147/prbm.s220677.

4. Duckworth, Angela, and Stephanie Carlson. "Self-Regulation and School Success." 2013. repository.upenn.edu/cgi/viewcontent.cgi?article=1002&context=psychology_papers.

5. Bloch, Lian, Claudia M. Haase, and Robert W. Levenson. "Emotion Regulation Predicts Marital Satisfaction: More than a Wives' Tale." *Emotion* 14 (1): 130–44, 2014. doi.org/10.1037/a0034272; Gross, James J., and Oliver P. John. "Individual Differences in Two Emotion Regulation Processes: Implications for Affect, Relationships, and Well-Being." *Journal of Personality and Social Psychology* 85 (2): 348–62, 2003. doi.org/10.1037/0022-3514.85.2.348.

6. Vogel, S., and L. Schwabe. "Learning and Memory Under Stress: Implications for the Classroom." *NPJ Science of Learning* 1 (16011), 2016. doi.org/10.1038/npjscilearn.2016.11.

7. Diamond, L. M., and L. G. Aspinwall. "Emotion Regulation Across the Life Span: An Integrative Perspective Emphasizing Self-Regulation, Positive Affect, and Dyadic Processes." *Motivation and Emotion* 27 (2), 125–56, 2003. doi.org/10.1023/A:1024521920068.

8. Nakagawa, Takeshi, et al. "Age, Emotion Regulation, and Affect in Adulthood: The Mediating Role of Cognitive Reappraisal." *Japanese Psychological Research* 59 (4): 301–8, 2017. doi.org/10.1111/jpr.12159; Silvers, Jennifer A., et al. "Age-Related Differences in Emotional Reactivity, Regulation, and Rejection Sensitivity in Adolescence." *Emotion* 12 (6): 1235–47, 2012. doi.org/10.1037/a0028297.

9. Lantrip and Huang. "Cognitive Control."

10. Chowdhury, Madhuleena Roy. "What Is Emotion Regulation? + 6 Emotional Skills and Strategies." PositivePsychology.com. August 13, 2019. positivepsychology.com/emotion-regulation/; Van Bockstaele, Bram, et al. "Choose Change: Situation Modification, Distraction, and Reappraisal in Mild Versus Intense Negative Situations." *Motivation and Emotion* 44, 2019. doi.org/10.1007/s11031-019-09811-8.

11. Ochsner, Kevin N., Jennifer A. Silvers, and Jason T. Buhle. "Functional Imaging Studies of Emotion Regulation: A Synthetic Review and Evolving Model of the Cognitive Control of Emotion." *Annals of the New York Academy of Sciences* 1251 (1): E1–24, 2012. doi.org/10.1111/j.1749-6632.2012.06751.x.

12. Paschke, Lena M., et al. "Individual Differences in Self-Reported Self-Control Predict Successful Emotion Regulation." *Social Cognitive and Affective Neuroscience* 11 (8): 1193–204, 2016. doi.org/10.1093/scan/nsw036.

13. Paschke, et al. "Individual Differences."

14. Harvard Health Publishing. "Understanding the Stress Response." Harvard Health, May 1, 2018. health.harvard.edu/staying-healthy/understanding-the-stress-response.

15. Aristizabal, Juan-Pablo, et al. "Use of Heart Rate Variability Biofeedback to Reduce the Psychological Burden of Frontline Healthcare Professionals against COVID-19." *Frontiers in Psychology* 11 (October), 2020. doi.org/10.3389/fpsyg.2020.572191.

16. Mather, Mara, and Julian F. Thayer. "How Heart Rate Variability Affects Emotion Regulation Brain Networks." *Current Opinion in Behavioral Sciences* 19 (February): 98–104, 2018. doi.org/10.1016/j.cobeha.2017.12.017; Appelhans, Bradley M., and Linda J. Luecken. "Heart Rate Variability as an Index of Regulated Emotional Responding." *Review of General Psychology* 10 (3): 229–40, 2006. doi.org/10.1037/1089-2680.10.3.229.

17. Kaufman, Erin A., et al. "The Difficulties in Emotion Regulation Scale Short Form (DERS-SF): Validation and Replication in Adolescent and Adult Samples." *Journal of Psychopathology and Behavioral Assessment* 38 (3): 443–55, 2015. doi.org/10.1007/s10862-015-9529-3.

18. Note that the MFA assesses feelings found to be associated with happiness in positive psychology (gratitude, awe, compassion, etc.) as well as more general feelings, such as those measured by the widely used Positive and Negative Affect Schedule survey; Watson, David, Lee Anna Clark, and Auke Tellegen. "Development and Validation of Brief Measures of Positive and Negative Affect: The PANAS Scales." *Journal of Personality and Social Psychology* 54 (6): 1063–70, 1988. doi .org/10.1037/0022-3514.54.6.1063.

Chapter 9: Memory and Learning

1. "Scientific Reports on Highly Superior Autobiographical Memory." n.d. Center for the Neurobiology of Learning and Memory. cnlm.uci.edu/hsam/scientific-reports/.

2. Patihis, L., et al. "False Memories in Highly Superior Autobiographical Memory Individuals." *Proceedings of the National Academy of Sciences* 110 (52): 20947–52, 2013. doi.org/10.1073 /pnas.1314373110.

3. Veiseh, Nima. "TEDxRhodes: Memory and Mindfulness." December 9, 2016. youtube.com /watch?v=9rR0VHBUY0A&feature=emb_logo.

4. "Neuroscience: How Much Power, in Watts, Does the Brain Use?" n.d. Psychology & Neuroscience Stack Exchange. psychology.stackexchange.com/questions/12385/how-much-power-in-watts -does-the-brain-use; "Computation Power: Human Brain Versus Supercomputer." Foglets: Science Discovery. April 10, 2019. foglets.com/supercomputer-vs-human-brain/#:~:text=The%20 amount%20of%20energy%20required.

5. Patihis, et al. "False Memories."

6. LePort, Aurora K. R., et al. "Highly Superior Autobiographical Memory: Quality and Quantity of Retention over Time." *Frontiers in Psychology* (6), 2016. doi.org/10.3389/fpsyg.2015.02017.

7. Urcelay, G. P., and R. R. Miller. "Retrieval from Memory." *Learning and Memory: A Comprehensive Reference* 1: 53–73, 2008. doi.org/10.1016/b978-012370509-9.00075-9; Squire, Larry R., et al. "Memory Consolidation." *Cold Spring Harbor Perspectives in Biology* 7 (8): a021766, 2015. doi.org/10.1101/cshperspect.a021766.

8. Hockley, William E., Fahad N. Ahmad, and Rosemary Nicholson. "Intentional and Incidental Encoding of Item and Associative Information in the Directed Forgetting Procedure." *Memory & Cognition* 44 (2): 220–28, 2015. doi.org/10.3758/s13421-015-0557-8.

9. Bauer, Patricia J., et al. "Neural Correlates of Autobiographical Memory Retrieval in Children and Adults." *Memory* 25 (4): 450–66, 2016. doi.org/10.1080/09658211.2016.1186699.

10. When something is new, those who recruit PFC the most usually perform the best. (Duncan and Owen 2000, Poldrack, et al. 2005.) But when you are really good at it, you are NOT using PFC as much. (Chein and Schneider 2005; Garavan, et al. 2000; Landau, et al. 2007; Milham, et al. 2003; Miller, et al. 2003.) From slide 50 of pdx.edu/social-determinants-health/sites/www.pdx .edu.social-determinants-health/files/Adele%20Diamond.pdf.

11. Fine Sasse, Stephanie. *Learning and Memory: Frontal Cortex, Striatum, Amygdala, Cerebellum, Hippocampus.* 2021.

12. Blumenstyk, Goldie. "By 2020, They Said, 2 out of 3 Jobs Would Need More Than a High-School Diploma. Were They Right?" *Chronicle of Higher Education.* January 22, 2020. chronicle .com/newsletter/the-edge/2020-01-22.

13. Gumbel, Peter. "How Will Automation Affect Jobs, Skills, and Wages?" Podcast. McKinsey Global Institute. 2018.

14. Bourke, Juliet. "The Overwhelmed Employee: Simplify the Work Environment." Deloitte Australia. October 2014. deloitte.com/au/en/pages/human-capital/articles/overwhelmed-employee -simplify-environment.html.

15. Center for the Neurobiology of Learning and Memory. "Highly Superior Autobiographical Memory." 2017. cnlm.uci.edu/hsam/.

16. Lapp, Danielle C. *Don't Forget!: Easy Exercises for a Better Memory at Any Age.* New York: McGraw-Hill, 1987.

Chapter 10: Creativity

1. Wujec, Tom. "TED: Build a Tower, Build a Team." February 2010. ted.com/talks/tom_wujec_build_a_tower_build_a_team?language=en.

2. Jauk, Emanuel, et al. "The Relationship between Intelligence and Creativity: New Support for the Threshold Hypothesis by Means of Empirical Breakpoint Detection." *Intelligence* 41(4), 212–21, 2013. doi.org/10.1016/j.intell.2013.03.003.

3. Kaufman, Alan S. *IQ Testing 101.* New York: Springer, 2009.

4. Jauk, et al. "The Relationship between Intelligence and Creativity."

5. Adapted from the University of Kent, "Mathematical Lateral Logic Test." n.d. kent.ac.uk/ces/tests/maths-logic-test.html.

6. King, Stephen. "Stephen King: Can a Novelist Be Too Productive?" *New York Times.* August 27, 2015. nytimes.com/2015/08/31/opinion/stephen-king-can-a-novelist-be-too-productive.html.

7. Simonton, Dean Keith. "Creative Productivity: A Predictive and Explanatory Model of Career Trajectories and Landmarks." *Psychological Review* 104 (1): 66–89, 1997. doi.org/10.1037/0033-295x.104.1.66.

8. Yamada, Yohei, and Masayoshi Nagai. "Positive Mood Enhances Divergent but Not Convergent Thinking." *Japanese Psychological Research* 57 (4): 281–87, 2015. doi.org/10.1111/jpr.12093.

9. Kaufman, Scott Barry. "The Emotions That Make Us More Creative." *Harvard Business Review.* August 12, 2015. hbr.org/2015/08/the-emotions-that-make-us-more-creative.

10. Kaufman, Scott Barry. "The Emotions."

11. Baer, John. "Creativity Doesn't Develop in a Vacuum." *New Directions for Child and Adolescent Development* (151): 9–20, 2016. doi.org/10.1002/cad.20151.

12. Kaufman, Scott Barry. "The Real Neuroscience of Creativity." *Scientific American.* August 19, 2013. blogs.scientificamerican.com/beautiful-minds/the-real-neuroscience-of-creativity/.

13. Beaty, Roger E., et al. "Robust Prediction of Individual Creative Ability from Brain Functional Connectivity." *Proceedings of the National Academy of Sciences* 115 (5): 1087–92, 2018. doi.org/10.1073/pnas.1713532115.

14. Liu, Siyuan, et al. "Neural Correlates of Lyrical Improvisation: An FMRI Study of Freestyle Rap." *Scientific Reports* 2 (1), 2012. doi.org/10.1038/srep00834; Limb, Charles J., and Allen R. Braun. "Neural Substrates of Spontaneous Musical Performance: An FMRI Study of Jazz Improvisation." *PLOS One* 3 (2): e1679, 2018. doi.org/10.1371/journal.pone.0001679.

15. Csikszentmihalyi, Mihaly. *Creativity: The Psychology of Discovery and Invention.* New York, London: Harper Perennial Modern Classics, 2013.

16. Carson, Shelley H., Jordan B. Peterson, and Daniel M. Higgins. "Reliability, Validity, and Factor Structure of the Creative Achievement Questionnaire." *Creativity Research Journal* 17 (1): 37–50, 2015. doi.org/10.1207/s15326934crj1701_4.

Chapter 11: Choosing Your Mental Target

1. Three skyline images, pages 115–16: iStock. *New York City Skyline Silhouette Vector Illustration.* https://www.istockphoto.com/vector/new-york-city-skyline-silhouette-vector-illustration-gm1152078220-312436161.

2. Hough, Lory. "Beyond Average." Harvard Graduate School of Education. 2015. gse.harvard
   .edu/news/ed/15/08/beyond-average.

3. "Twice-Exceptional Students." Nagc.org, 2000. nagc.org/resources-publications/resources
   -parents/twice-exceptional-students.

4. Schaefer, Charles E., and Howard L. Millman. *How to Help Children with Common Problems.*
   Northvale, NJ: J. Aronson, 1994.

5. Keller, Arielle S., et al. "Paying Attention to Attention in Depression." *Translational Psychiatry* 9
   (1), 2019. doi.org/10.1038/s41398-019-0616-1.

6. Pacheco-Unguetti, Antonia Pilar, et al. "Attention and Anxiety." *Psychological Science* 21 (2):
   298–304, 2010. doi.org/10.1177/0956797609359624.

7. "Sleep, Learning, and Memory." Harvard.edu, 2019. healthysleep.med.harvard.edu/healthy
   /matters/benefits-of-sleep/learning-memory.

8. Lucius, Khara. "'Brain Fog': Exploring a Symptom Commonly Encountered in Clinical Practice." *Alternative and Complementary Therapies* 27 (1): 23–30, 2021. doi.org/10.1089/act.2020
   .29313.klu.

9. Centers for Disease Control and Prevention. "What Is ADHD?" September 19, 2018. cdc.gov
   /ncbddd/adhd/facts.html.

10. Ram, Nilam, et al. "Cognitive Performance Inconsistency: Intraindividual Change and Vari-
    ability." *Psychology and Aging* 20 (4): 623–33, 2005. doi.org/10.1037/0882-7974.20.4.623.

11. Facer-Childs, Elise R., Sophie Boiling, and George M. Balanos. "The Effects of Time of Day
    and Chronotype on Cognitive and Physical Performance in Healthy Volunteers." *Sports Medi-
    cine Open* 4 (1), October 24, 2018. doi.org/10.1186/s40798-018-0162-z.

12. Smarr, Benjamin L., and Aaron E. Schirmer. "3.4 Million Real-World Learning Management
    System Logins Reveal the Majority of Students Experience Social Jet Lag Correlated with
    Decreased Performance." *Scientific Reports* 8 (1), 2018. doi.org/10.1038/s41598-018-23044-8.

13. Facer-Childs, et al. "The Effects of Time of Day."

14. Dunster, Gideon P., et al. "Sleepmore in Seattle: Later School Start Times Are Associated with
    More Sleep and Better Performance in High School Students." *Science Advances* 4 (12):
    eaau6200, 2018. doi.org/10.1126/sciadv.aau6200.

15. Tamnes, C. K., et al. "Becoming Consistent: Developmental Reductions in Intraindividual
    Variability in Reaction Time Are Related to White Matter Integrity." *Journal of Neuroscience* 32
    (3): 972–82, 2012. doi.org/10.1523/jneurosci.4779-11.2012; Ram, Nilam, et al. "Cognitive Per-
    formance Inconsistency." 2005.

16. Gamaldo, Alyssa A., et al. "Variability in Performance: Identifying Early Signs of Future Cogni-
    tive Impairment." *Neuropsychology* 26 (4): 534–40, 2012. doi.org/10.1037/a0028686

17. Kiziltas, Semiha, Burcu Akinci, and Cleotilde Gonzalez. "Comparison of Experienced and
    Novice Cost Estimator Behaviors in Information Pull and Push Methods." *Canadian Journal of
    Civil Engineering* 37 (2): 290–301, 2010. doi.org/10.1139/l09-152.

Chapter 12: Life Scoring

1. Doran, G. T. "There's a S.M.A.R.T. Way to Write Management's Goals and Objectives." *Management Review.* 70 (11): 35–36, 1981.

2. Quantified Self. "Show & Tell: Tracking What I Do Versus What I Say I'll Do." n.d. quantifiedself
   .com/show-and-tell/?project=1097.

3. "Learn to Human Better." n.d. inneru.coach/.

4. "Designing Your Life." 2007. ocw.mit.edu/courses/athletics-physical-education-and-recreation
   /pe-550-designing-your-life-january-iap-2007/assignments/assign01.pdf.

5. "The Wheel of Life." Positive Psychology. 2016. positivepsychology.com/wp-content/uploads /2016/11/The-Wheel-of-Life.pdf.

6. Pascha, Mariana. 2019. "The PERMA Model: Your Scientific Theory of Happiness." Positive Psychology. July 3, 2019. positivepsychology.com/perma-model/.

7. "Ed Diener, Subjective Well-Being." n.d. labs.psychology.illinois.edu/~ediener/SWLS.html #:~:text=The%20SWLS%20is%20a%20short.

8. Adapted from: 18 Areas of Life, Handel Method: "PE.550 Designing Your Life." 2007. ocw.mit .edu/courses/athletics-physical-education-and-recreation/pe-550-designing-your-life-january -iap-2007/assignments/assign01.pdf; Diener, E., et al. "The Satisfaction with Life Scale." *Journal of Personality Assessment* 49 (1): 71–75, 1985. doi.org/10.1207/s15327752jpa4901_13; positive psychology.com/wp-content/uploads/2016/11/The-Wheel-of-Life.pdf.

Chapter 13: Placebo on Purpose

1. Kopp, Vincent. *Henry K. Beecher, M.D.: Contrarian (1904–1976)*. American Society of Anesthesiologists, 1999. web.archive.org/web/20001119013500/http:/www.asahq.org/NEWSLET TERS/1999/09_99/beecher0999.html; some controversy over this story exists: shannonharvey .com/blogs/blog/this-is-why-you-shouldnt-believe-everything-you-read-about-your-health.

2. "Placebos Are Getting More Effective. Drugmakers Are Desperate to Know Why." *Wired*. August 24, 2009. www.wired.com/2009/08/ff-placebo-effect/.

3. Beecher, Henry K. "The Powerful Placebo." *Journal of the American Medical Association* 159 (17): 1602, 1955. doi:10.1001/jama.1955.02960340022006.

4. Levine, Jon D., Newton C. Gordon, and Howard L. Fields. "The Mechanism of Placebo Analgesia." *The Lancet* 312 (8091): 654–57, 1978. doi:10.1016/s0140-6736(78)92762-9; Lipman, Jonathan J., et al. "Peak B Endorphin Concentration in Cerebrospinal Fluid: Reduced in Chronic Pain Patients and Increased during the Placebo Response." *Psychopharmacology* 102 (1): 112–16, 1990. doi:10.1007/bf02245754; Vachon-Presseau, Etienne, et al. "Brain and Psychological Determinants of Placebo Pill Response in Chronic Pain Patients." *Nature Communications* 9 (1), 2018. doi:10.1038/s41467-018-05859-1; "Sugar Pills Relieve Pain for Chronic Pain Patients." *ScienceDaily*. Northwestern University. September 12, 2018. sciencedaily.com/releases/2018 /09/180912133542.htm.

5. Pittrof, Rudiger. "Placebo Treatment in Mild to Moderate Depression." *British Journal of General Practice* 61 (584), 2011. doi:10.3399/bjgp11x561285.

6. Darragh, Margot, et al. "A Take-Home Placebo Treatment Can Reduce Stress, Anxiety and Symptoms of Depression in a Non-Patient Population." *Australian & New Zealand Journal of Psychiatry* 50 (9): 858–65, 2016. doi:10.1177/0004867415621390.

7. Fuente-Fernández, R. de la, S. Lidstone, and A. J. Stoessl. "Placebo Effect and Dopamine Release." *Parkinson's Disease and Related Disorders* 70: 415–18, 2006. doi:10.1007/978-3-211-45295-0_62.

8. Sihvonen, Raine, et al. "Arthroscopic Partial Meniscectomy Versus Sham Surgery for a Degenerative Meniscal Tear." *New England Journal of Medicine* 369 (26): 2515–24, 2013. doi:10.1056/ nejmoa1305189; Talbot, Margaret. "The Placebo Prescription." *New York Times*. January 9, 2000. nytimes.com/2000/01/09/magazine/the-placebo-prescription.html; Moseley, J. Bruce, et al. "A Controlled Trial of Arthroscopic Surgery for Osteoarthritis of the Knee." *New England Journal of Medicine* 347 (2): 81–88, 2002. doi:10.1056/nejmoa013259.

9. Wadyka, Sally. "3 Ways to Use the Placebo Effect to Have a Better Day." Health.com. 2021. health.com/mind-body/3-ways-to-use-the-placebo-effect-to-have-a-better-day; Draganich, Christina, and Kristi Erdal. "Placebo Sleep Affects Cognitive Functioning." *Journal of Experimental Psychology: Learning, Memory, and Cognition* 40 (3): 857–64, 2014. doi:10.1037/a0035546.

10. Kross, E., et al. "Social Rejection Shares Somatosensory Representations with Physical Pain." *Proceedings of the National Academy of Sciences* 108 (15): 6270–75, 2011. doi:10.1073/pnas.1102693108; Szalavitz, Maia. "New Test Distinguishes Physical From Emotional Pain in Brain for First Time." *Time*. May 6, 2013. healthland.time.com/2013/05/06/a-pain-detector-for-the-brain/.

11. Graham, Sarah. "Brain's Own Pain Relievers at Work in Placebo Effect, Study Suggests." *Scientific American*. August 24, 2005. scientificamerican.com/article/brains-own-pain-relievers/.

12. Hall, Kathryn T., Joseph Loscalzo, and Ted J. Kaptchuk. "Genetics and the Placebo Effect: the Placebome." *Trends in Molecular Medicine* 21 (5): 285–94, 2015. doi:10.1016/j.molmed .2015.02.009.

13. "Parkinson's Disease." Mayo Foundation for Medical Education and Research. December 8, 2020. mayoclinic.org/diseases-conditions/parkinsons-disease/symptoms-causes/syc-20376055; Triarhou, Lazaros C. "Dopamine and Parkinson's Disease." National Center for Biotechnology Information. U.S. National Library of Medicine. January 1, 1970. ncbi.nlm.nih.gov/books/NBK6271.

14. Hall, et al. "Genetics and the Placebo Effect: the Placebome."

15. Ongaro, Giulio, and Ted J. Kaptchuk. "Symptom Perception, Placebo Effects, and the Bayesian Brain." *Pain* 1, 2018. doi:10.1097/00006396-900000000-98882.

16. "You Get What You Pay For? Costly Placebo Works Better Than Cheap One." ScienceDaily. March 5, 2008. sciencedaily.com/releases/2008/03/080304173339.htm.

17. "Placebos Are Getting More Effective"; Waber R. L., et al. "Commercial Features of Placebo and Therapeutic Efficacy." *Jama* 299 (9): 1016, 2008. doi:10.1001/jama.299.9.1016.

18. Adam, Hajo, and Adam D. Galinsky. "Enclothed Cognition." *Journal of Experimental Social Psychology* 48 (4): 918–25, 2012. doi:10.1016/j.jesp.2012.02.008.

19. Dispenza, Joe. *You Are the Placebo: Making Your Mind Matter*. Carlsbad, CA: Hay House, 2015.

20. Madrigal, Alexis C. "The Dark Side of the Placebo Effect: When Intense Belief Kills." *The Atlantic*. 2011. theatlantic.com/health/archive/2011/09/the-dark-side-of-the-placebo-effect -when-intense-belief-kills/245065/.

21. Jakšić, Nenad, Branka Aukst-Margetić, and Miro Jakovljević. "Does Personality Play a Relevant Role in the Placebo Effect?" *Psychiatria Danubina*. U.S. National Library of Medicine, 2013. March. ncbi.nlm.nih.gov/pubmed/23470602.

22. Vachon-Presseau, Etienne, et al. "Brain and Psychological Determinants of Placebo Pill Response in Chronic Pain Patients." *Nature Communications* 9 (1), 2018. doi:10.1038 /s41467-018-05859-1.

23. Romm, Cari. "Is the Placebo Effect in Your DNA?" *The Atlantic*, April 13, 2015. theatlantic .com/health/archive/2015/04/is-the-placebo-effect-in-your-dna/390360/.

24. "rs4680." SNPedia. 2021. snpedia.com/index.php/Rs4680; Hall, et al. "Genetics and the Placebo Effect: the Placebome."

25. Kaptchuk, Ted J., et al. "Placebos without Deception: A Randomized Controlled Trial in Irritable Bowel Syndrome." *PLOS One* 5 (12), 2010. doi:10.1371/journal.pone.0015591.

26. Locher, Cosima, et al. "Is the Rationale More Important than Deception? A Randomized Controlled Trial of Open-Label Placebo Analgesia." *Pain* 158 (12): 2320–28, 2017. doi:10.1097/j .pain.0000000000001012.

27. "What Oprah Learned from Jim Carrey." Oprah.com. October 12, 2011. oprah.com/oprahs -lifeclass/what-oprah-learned-from-jim-carrey-video.

28. "Natan Sharansky: How Chess Kept One Man Sane." BBC News. January 3, 2014. bbc.com /news/magazine-25560162.

29. Schmemann, Serge. "Kasparov Beaten in Israel, by Russians." *New York Times*. October 16, 1996. nytimes.com/1996/10/16/world/kasparov-beaten-in-israel-by-russians.html.

30. Pascual-Leone, A., et al. "Modulation of Muscle Responses Evoked by Transcranial Magnetic Stimulation during the Acquisition of New Fine Motor Skills." *Journal of Neurophysiology* 74 (3): 1037–45, 1995. doi:10.1152/jn.1995.74.3.1037.

31. Denis, M. "Visual Imagery and the Use of Mental Practice in the Development of Motor Skills." *Canadian Journal of Applied Sport Sciences.* U.S. National Library of Medicine. 2021. pubmed .ncbi.nlm.nih.gov/3910301/; Ietswaart, Magdalena, et al. "Mental Practice with Motor Imagery in Stroke Recovery: Randomized Controlled Trial of Efficacy." *Brain* 134 (5): 1373–86, 2011. doi:10.1093/brain/awr077; Page, Stephen J., Peter Levine, and Anthony Leonard. "Mental Practice in Chronic Stroke." *Stroke* 38 (4): 1293–97, 2007. doi:10.1161/01.str.0000260205.67348.2b.

32. Special thanks goes to my sister for suggesting the word *provokes* for this sentence.

33. Schleider, Jessica, and John Weisz. "A Single-Session Growth Mindset Intervention for Adolescent Anxiety and Depression: 9-Month Outcomes of a Randomized Trial." *Journal of Child Psychology and Psychiatry* 59 (2): 160–70, 2017. doi:10.1111/jcpp.12811; "Effect Size (ES)." 2021. lbecker.uccs.edu/effect-size.

34. Rozenkrantz, Liron, et al. "Placebo Can Enhance Creativity." *PLOS One* 12 (9), 2017. doi:10.1371/journal.pone.0182466.

35. Guevarra, Darwin A., et al. "Placebos without Deception Reduce Self-Report and Neural Measures of Emotional Distress." *Nature Communications* 11 (1), 2020. doi:10.1038/s41467-020-17654-y; Schaefer, Michael, et al. "Open-Label Placebos Reduce Test Anxiety and Improve Self-Management Skills: A Randomized-Controlled Trial." *Scientific Reports* 9 (1), 2019. doi:10.1038/s41598-019-49466-6.

Chapter 14: Sweat

1. Diamond, Adele, and Daphne S. Ling. "Conclusions about Interventions, Programs, and Approaches for Improving Executive Functions That Appear Justified and Those That, despite Much Hype, Do Not." *Developmental Cognitive Neuroscience* 18: 34–48, 2016. doi:10.1016/j.dcn.2015.11.005.

2. Hillman, C. H., et al. "Effects of the FITKids Randomized Controlled Trial on Executive Control and Brain Function." *Pediatrics* 134 (4), 2014. doi:10.1542/peds.2013-3219.

3. Chang, Y. K., et al. "The Effects of Acute Exercise on Cognitive Performance: A Meta-Analysis." *Brain Research* 1453: 87–101, 2012. doi:10.1016/j.brainres.2012.02.068.

4. Based on Cohen's D of XX. Conversion from Becker, Dr. Lee A., "Effect Size Calculators." n.d. lbecker.uccs.edu/effect-size.

5. Randolph, Derek D., and Patrick J. O'Connor. "Stair Walking Is More Energizing than Low Dose Caffeine in Sleep Deprived Young Women." *Physiology & Behavior* 174: 128–35, 2017. doi:10.1016/j.physbeh.2017.03.013.

6. antoinedl.com/fichiers/public/ACSM-guidelines-2014.pdf.

7. Ratey, John J., and Eric Hagerman. *Spark! How Exercise Will Improve the Performance of Your Brain*. Quercus. 2010.

8. Westcott, Wayne L. "Resistance Training Is Medicine: Effects of Strength Training on Health." *Current Sports Medicine Reports* 11 (4): 209–16, 2012. doi:10.1249/jsr.0b013e31825dabb8.

9. Winter, Bernward, et al. "High Impact Running Improves Learning." *Neurobiology of Learning and Memory* 87 (4): 597–609, 2007. doi:10.1016/j.nlm.2006.11.003.

10. Van Dongen, Eelco V., et al. "Physical Exercise Performed Four Hours after Learning Improves

Memory Retention and Increases Hippocampal Pattern Similarity during Retrieval." *Current Biology* 26 (13): 1722–27, 2016. doi:10.1016/j.cub.2016.04.071.

11. Oaten, Megan, and Ken Cheng. "Longitudinal Gains in Self-Regulation from Regular Physical Exercise." *British Journal of Health Psychology* 11 (4): 717–33, 2006. doi:10.1348/135910706x96481.

12. Loy, Bryan D., Patrick J. O'Connor, and Rodney K. Dishman. "The Effect of a Single Bout of Exercise on Energy and Fatigue States: a Systematic Review and Meta-Analysis." *Fatigue: Biomedicine, Health & Behavior* 1 (4): 223–42, 2013. doi:10.1080/21641846.2013.843266; Randolph and O'Connor, "Stair Walking."

13. Edwards, Meghan K., et al. "Effects of Acute Aerobic Exercise or Meditation on Emotional Regulation." *Physiology & Behavior* 186: 16–24, 2018. doi:10.1016/j.physbeh.2017.12.037.

14. Association for Psychological Science. "A Positive Mood Allows Your Brain to Think More Creatively," December 15, 2010. psychologicalscience.org/news/releases/a-positive-mood-allows-your-brain-to-think-more-creatively.html; Zenasni, Franck, and Todd Lubart. "Effects of Mood States on Creativity." *Current Psychology Letters: Behaviour, Brain and Cognition* 2002/2, 8, 2002. doi:10.4000/cpl.205.

15. Steinberg, Hannah, et al. "Exercise Enhances Creativity Independently of Mood." *British Journal of Sports Medicine* 31 (3): 240–45, 1997. doi:10.1136/bjsm.31.3.240.

16. Oppezzo, Marily, and Daniel L. Schwartz. "Give Your Ideas Some Legs: The Positive Effect of Walking on Creative Thinking." *Journal of Experimental Psychology: Learning, Memory, and Cognition* 40 (4): 1142–52, 2014. doi:10.1037/a0036577.

17. Rominger, Christian, et al. "Creative Challenge: Regular Exercising Moderates the Association between Task-Related Heart Rate Variability Changes and Individual Differences in Originality." *PLOS One* 14 (7), 2019. doi:10.1371/journal.pone.0220205.

Chapter 15: Let There Be (Blue) Light

1. health.harvard.edu/blog/seasonal-affective-disorder-bring-on-the-light-201212215663.

2. Vandewalle, Gilles, Pierre Maquet, and Derk-Jan Dijk. "Light as a Modulator of Cognitive Brain Function." *Trends in Cognitive Sciences* 13 (10): 429–38, 2009. doi.org/10.1016/j.tics.2009.07.004.

3. Alkozei, Anna, et al. "Acute Exposure to Blue Wavelength Light during Memory Consolidation Improves Verbal Memory Performance." *PLOS One* 12 (9), 2017. doi.org/10.1371/journal.pone.0184884.

4. Ma, Zhiqiang, Yang Yang, Chongxi Fan, et al. "Melatonin as a Potential Anticarcinogen for Non-Small-Cell Lung Cancer." *Oncotarget* 7 (29): 46768–84, 2016. doi:10.18632/oncotarget.8776. ncbi.nlm.nih.gov/pmc/articles/PMC5216835/. Licensed under the Creative Commons Attribution 4.0 International license.

5. Viola, Antoine U., et al. "Blue-Enriched White Light in the Workplace Improves Self-Reported Alertness, Performance and Sleep Quality." *Scandinavian Journal of Work, Environment & Health* 34 (4): 297–306, 2008. doi.org/10.5271/sjweh.1268.

6. Craig, Michael. "Seasonal Affective Disorder: Bring on the Light," Harvard Health. December 21, 2012. health.harvard.edu/blog/seasonal-affective-disorder-bring-on-the-light-201212215663.

7. Taillard, Jacques, et al. "In-Car Nocturnal Blue Light Exposure Improves Motorway Driving: A Randomized Controlled Trial." *PLOS One* 7 (10): e46750, 2012. doi.org/10.1371/journal.pone.0046750.

8. Koninklijke Philips NV. "Buy the Philips GoLITE BLU Energy Light HF3422/60 Energy Light." n.d. usa.philips.com/c-p/HF3422_60/golite-blu-energy-light/overview#see-all-benefits.

9. Viola, et al. "Blue-Enriched White Light."

10. Beaven, C. Martyn, and Johan Ekström. "A Comparison of Blue Light and Caffeine Effects on Cognitive Function and Alertness in Humans." *PLOS One* 8 (10): e76707, 2013. doi.org/10.1371 /journal.pone.0076707.

11. Duke Health. "Myth or Fact: People with Light Eyes Are More Sensitive to Sunlight." August 27, 2013. dukehealth.org/blog/myth-or-fact-people-light-eyes-are-more-sensitive-sunlight.

12. Alkozei, Anna, et al. "Acute Exposure to Blue Wavelength Light during Memory Consolidation Improves Verbal Memory Performance." *PLOS One* 12 (9), 2017. doi.org/10.1371/journal .pone.0184884.

13. Alkozei, et al., "Acute Exposure."

Chapter 16: Rewriting the Brain's Signature

1. Bindrā, Abhinava, and Rohit Brijnath. *A Shot at History: My Obsessive Journey to Olympic Gold.* Uttar Pradesh, India: Harper Sport. 2013.

2. Bindrā and Brijnath, *A Shot at History.*

3. "Meditation Dramatically Changes Body Temperatures." *Harvard Gazette*, April 18, 2002. news.harvard.edu/gazette/story/2002/04/meditation-dramatically-changes-body -temperatures/.

4. Krol, Laurens R. *Ten Seconds of Simulated EEG Data in the Five Differently Named Frequency Bands of Neural Oscillations, or Brainwaves: Delta, Theta, Alpha, Beta, and Gamma.* December 3, 2020. en.wikipedia.org/wiki/File:EEG_Brainwaves.svg.

5. "Clarification of Neurofeedback." International Society for Neuroregulation and Research. February 6, 2017. isnr.org/in-defense-of-neurofeedback.

6. psychcentral.com/blog/neurofeedback-therapy-an-effective-non-drug-treatment-for-adhd/.

7. neurodevelopmentcenter.com/psychological-disorders/adhd-add-symptoms/adhd-treatment -without-medication/.

8. potencialmenteacademia.com.br/wp-content/uploads/2018/11/gruzelier2014-EEG -neurofeedback-for-optimising-performance.-I_-.pdf.

9. Marzbani, Marateb, and Mansourian. "Methodological Note."

10. University of Goldsmiths London. "Mind Control Can Make You A Better Surgeon." Science-Daily. August 20, 2009. sciencedaily.com/releases/2009/08/090819125319.htm.

11. Association for Applied Biopsychology and Biofeedback. n.d. "Review of New Research Regarding EEG Neurofeedback and Musicians." aapb.org/i4a/pages/index.cfm?pageID=3388.

12. Murphy, Jen. "Mental, Physical Training for Olympic Volleyball." *Wall Street Journal,* July 11, 2012. wsj.com/articles/SB10001424052702304141204577510740730336270.

13. McAllister, Mike. "How Bryson Trains His Brain." PGATour.com. February 19, 2019. pgatour .com/long-form/2019/02/19/bryson-dechambeau-brain-training.html.

14. Bhayee, Sheffy, et al. "Attentional and Affective Consequences of Technology Supported Mindfulness Training: A Randomised, Active Control, Efficacy Trial." *BMC Psychology* 4, 2016. doi .org/10.1186/s40359-016-0168-6.

15. Goddard, Nick. "Psychology." *Core Psychiatry* 63–82, 2012. doi.org/10.1016/b978-0-7020 -3397-1.00005-7.

16. "About Brian—Brian Othmer Foundation." n.d. brianothmerfoundation.org/about-brian/.

17. Fleischman, Matthew J., and Siegfried Othmer. "Case Study: Improvements in IQ Score and Maintenance of Gains Following EEG Biofeedback with Mildly Developmentally Delayed Twins." *Journal of Neurotherapy* 9 (4): 35–46, 2006. doi.org/10.1300/j184v09n04_03.

18. Othmer, Siegfried. Email to Elizabeth Ricker, March 15, 2021.

19. Engelbregt, H. J., et al. "Short- and Long-Term Effects of Sham-Controlled Prefrontal

EEG-Neurofeedback Training in Healthy Subjects." *Clinical Neurophysiology* 127 (4): 1931–37, 2016. doi.org/10.1016/j.clinph.2016.01.004.

20. forum.choosemuse.com/t/measuring-the-latency-of-sending-data-from-muse-muse-monitor/1990/7.

21. Meinlschmidt, Gunther, et al. "Smartphone-Based Psychotherapeutic Micro-Interventions to Improve Mood in a Real-World Setting." *Frontiers in Psychology* 7: 1112, 2016. doi.org/10.3389/fpsyg.2016.01112.

22. "About Dr. Browne." n.d. Neurenics Psychology, Inc. neurenics.com/about/.

Chapter 17: Serious Games

1. Granic, Isabela, Adam Lobel, and Rutger C. M. E. Engels. "The Benefits of Playing Video Games." *American Psychologist* 69 (1): 66, 2014.

2. Azadegan, Aida, Johann C. K. H. Riedel, and Jannicke Baalsrud Hauge. "Serious Games Adoption in Corporate Training." In *International Conference on Serious Games Development and Applications*. Berlin and Heidelberg: Springer, 2012. 74–85.

3. Csikszentmihalyi, Mihaly. *Finding Flow: The Psychology of Engagement with Everyday Life*. New York: Basic Books, 2008.

4. Public domain.

5. Wise, Roy A. "Dopamine, Learning and Motivation." *Nature Reviews Neuroscience* 5 (6): 483–94, 2004. doi.org/10.1038/nrn1406.

6. Freeman, Bob. "Researchers Examine Video Gaming's Benefits." U.S. Air Force. n.d. af.mil/News/Article-Display/Article/117856/researchers-examine-video-gamings-benefits/.

7. Palaus, Marc, et al. "Neural Basis of Video Gaming: A Systematic Review." *Frontiers in Human Neuroscience* 11, 2017. doi.org/10.3389/fnhum.2017.00248.

8. BioMed Central Limited. "Is Tetris Good for the Brain?" ScienceDaily. sciencedaily.com/releases/2009/09/090901082851.htm

9. Shute, Valerie J., Matthew Ventura, and Fengfeng Ke. "The Power of Play: The Effects of Portal 2 and Lumosity on Cognitive and Noncognitive Skills." *Computers & Education* 80: 58–67, 2015. doi.org/10.1016/j.compedu.2014.08.013.

10. Strenziok, Maren, et al. "Neurocognitive Enhancement in Older Adults: Comparison of Three Cognitive Training Tasks to Test a Hypothesis of Training Transfer in Brain Connectivity." *NeuroImage* 85 Pt 3: 1027–39, 2014. doi.org/10.1016/j.neuroimage.2013.07.069.

11. Tennstedt, Sharon L., and Frederick W. Unverzagt. "The ACTIVE Study." *Journal of Aging and Health* 25 (8 suppl): 3S20S, 2013. doi.org/10.1177/0898264313518133.

12. Owsley, Cynthia, et al. "Timed Instrumental Activities of Daily Living Tasks: Relationship to Cognitive Function and Everyday Performance Assessments in Older Adults." *Gerontology* 48 (4): 254–65, 2002. doi.org/10.1159/000058360.

13. Tennstedt and Unverzagt, "The ACTIVE Study."

14. Wolinsky, Fredric D., et al. "A Randomized Controlled Trial of Cognitive Training Using a Visual Speed of Processing Intervention in Middle Aged and Older Adults." *PLOS One* 8 (5): e61624, 2013. doi.org/10.1371/journal.pone.0061624.

15. Hardy, Joseph L., et al. "Enhancing Cognitive Abilities with Comprehensive Training: A Large, Online, Randomized, Active-Controlled Trial." *PLOS One* 10 (9): e0134467, 2015. doi.org/10.1371/journal.pone.0134467.

16. Anguera, J. A., et al. "Video Game Training Enhances Cognitive Control in Older Adults." *Nature* 501 (7465): 97–101, 2013. doi.org/10.1038/nature12486.

17. Kühn, Simone, et al. "Does Playing Violent Video Games Cause Aggression? A Longitudinal Intervention Study." *Molecular Psychiatry* 24, 2018. doi.org/10.1038/s41380-018-0031-7.

18. Prescott, Anna T., James D. Sargent, and Jay G. Hull. "Metaanalysis of the Relationship between Violent Video Game Play and Physical Aggression over Time." *Proceedings of the National Academy of Sciences* 115 (40): 9882–88, 2018. doi.org/10.1073/pnas.1611617114.

19. "Mortality and Morbidity Statistics." n.d. icd.who.int/browse11/l-m/en#/http://id.who.int/icd/entity/1448597234.

20. "Internet Gaming." Psychiatry.org, 2013. psychiatry.org/patients-families/internet-gaming.

21. Przybylski, Andrew K., Netta Weinstein, and Kou Murayama. "Internet Gaming Disorder: Investigating the Clinical Relevance of a New Phenomenon." *American Journal of Psychiatry* 174 (3): 230–36, 2017. doi.org/10.1176/appi.ajp.2016.16020224.

22. González-Bueso, Vega, et al. "Association Between Internet Gaming Disorder or Pathological Video-Game Use and Comorbid Psychopathology: A Comprehensive Review." *International Journal of Environmental Research and Public Health* 15 (4): 668, 2018. doi.org/10.3390/ijerph15040668.

23. González-Bueso, et al. "Association Between Internet Gaming Disorder"; Przybylski, Weinstein, and Murayama. "Internet Gaming Disorder."

24. Palaus, Marc, et al. "Neural Basis of Video Gaming: A Systematic Review." *Frontiers in Human Neuroscience* 11, 2017. doi.org/10.3389/fnhum.2017.00248.

25. Krishnan, Lavanya, et al. "Neural Strategies for Selective Attention Distinguish Fast-Action Video Game Players." *Brain Topography* 26 (1): 83–97, 2012. doi.org/10.1007/s10548-012-0232-3.

26. Green, C. Shawn, and Daphne Bavelier. "Action Video Game Modifies Visual Selective Attention." *Nature* 423 (6939): 534–37, 2003. doi.org/10.1038/nature01647.

27. Tsai, M.-H., R.-J. Cherng, and J.-Y. Chen, "Visuospatial Attention Abilities in the Action and Real Time Strategy Video Game Players as Compared with Nonplayers," in *2013 1st International Conference on Orange Technologies (ICOT)* (Tainan: IEEE), 264–265, 2013.

28. Jaeggi, Susanne M., et al. "Improving Fluid Intelligence with Training on Working Memory." *Proceedings of the National Academy of Sciences* 105 (19): 6829–33, 2008.

29. Asprey, Dave. "The Father of Biohacking." blog.daveasprey.com/how-to-add-2-75-iq-points-per-hour-of-training/.

30. "HighIQPro IQ Increase Guarantee." highiqpro.com/high-iq-pro-training-guarantees.

31. Hemenover, Scott H., and Nicholas D. Bowman. "Video Games, Emotion, and Emotion Regulation: Expanding the Scope." *Annals of the International Communication Association* 42 (2): 125, n.d. academia.edu/37254137/Video_games_emotion_and_emotion_regulation_expanding_the_scope; Villani, Daniela, et al. "Videogames for Emotion Regulation: A Systematic Review." *Games for Health Journal* 7 (2): 85–99, 2018. doi.org/10.1089/g4h.2017.0108.

32. Iyadurai, L., et al. "Preventing Intrusive Memories after Trauma via a Brief Intervention Involving Tetris Computer Game Play in the Emergency Department: A Proof-of-Concept Randomized Controlled Trial." *Molecular Psychiatry* 23 (3): 674–82, 2017. doi.org/10.1038/mp.2017.23.

33. Villani, et al. "Videogames for Emotion Regulation."

34. Villani, et al. "Videogames for Emotion Regulation."

35. mightier.com/wp-content/uploads/2019/11/ScientificOverview.pdf.

36. Lobel, Adam, et al. "Designing and Utilizing Biofeedback Games for Emotion Regulation: The Case of Nevermind." In *Proceedings of the 2016 CHI Conference Extended Abstracts on Human Factors in Computing Systems,* ACM, 2016. 1945–1951.

37. McGonigal, Jane. "Transcript of 'The Game That Can Give You 10 Extra Years of Life.'"

TED, 2012. ted.com/talks/jane_mcgonigal_the_game_that_can_give_you_10_extra_years_of_life/transcript.

38. Bakker, David, et al. "Mental Health Smartphone Apps: Review and Evidence-Based Recommendations for Future Developments." *JMIR Mental Health* 3 (1): e7, 2016. doi.org/10.2196/mental.4984; Payne, Hannah E., Victor B. A. Moxley, and Elizabeth MacDonald. "Health Behavior Theory in Physical Activity Game Apps: A Content Analysis." *JMIR Serious Games* 3 (2): e4, 2015. doi.org/10.2196/games.4187; Roepke, Ann Marie, et al. "Randomized Controlled Trial of SuperBetter, a Smartphone-Based/Internet-Based Self-Help Tool to Reduce Depressive Symptoms." *Games for Health Journal* 4 (3): 235–46, 2015. doi.org/10.1089/g4h.2014.0046; Devan, Hemakumar, et al. "Evaluation of Self-Management Support Functions in Apps for People with Persistent Pain: Systematic Review." *JMIR MHealth and UHealth* 7 (2): e13080, 2019. doi.org/10.2196/13080.

39. Derlyatka, Anton, et al. "Bright Spots, Physical Activity Investments That Work: Sweatcoin: A Steps Generated Virtual Currency for Sustained Physical Activity Behaviour Change." *British Journal of Sports Medicine* 53 (18): 1195–96, 2019. doi.org/10.1136/bjsports-2018-099739.

40. Patoka, Josh. "14 Legit Apps That Will Pay You to Workout [*sic*]." Well Kept Wallet. April 19, 2020. wellkeptwallet.com/apps-pay-you-to-workout/.

41. Kelley, Paul, and Terry Whatson. "Making Long-Term Memories in Minutes: A Spaced Learning Pattern from Memory Research in Education. *Frontiers in Human Neuroscience* 7: 589, 2013. frontiersin.org/articles/10.3389/fnhum.2013.00589/full#h1.

42. Karpicke, Jeffrey D., and Althea Bauernschmidt. "Spaced Retrieval: Absolute Spacing Enhances Learning regardless of Relative Spacing." *Journal of Experimental Psychology: Learning, Memory, and Cognition* 37 (5): 1250–57, 2011. doi.org/10.1037/a0023436.

43. Kelley and Whatson. "Making Long-Term Memories in Minutes."

44. Icez. *Forgetting Curve with Spaced Repetition*, 2007. https://en.wikipedia.org/wiki/Forgetting_curve."

45. Karpicke, Jeffrey D., and J. R. Blunt. "Retrieval Practice Produces More Learning than Elaborative Studying with Concept Mapping." *Science* 331 (6018): 772–75, 2011. doi.org/10.1126/science.1199327.

46. Nakano, Dana. "Elevate Effectiveness Study Principal Author." 2015. elevateapp.com/assets/docs/elevate_effectiveness_october2015.pdf.

47. Grego, John. "Duolingo Effectiveness Study Final Report." 2012. static.duolingo.com/s3/DuolingoReport_Final.pdf.

48. Vesselinov, Roumen. "Measuring the Effectiveness of Rosetta Stone Final Report." 2009. resources.rosettastone.com/CDN/us/pdfs/Measuring_the_Effectiveness_RS-5.pdf.

49. "How Hockey Sense Training Changed the Game for USA Hockey." n.d. The Hockey IntelliGym. usahockeyintelligym.com/how-hockey-iq-training-changed-the-game-for-usa-hockey/.

50. "Concussion Study Indicates Hockey IntelliGym Training Could Reduce Chance of Injury." n.d. The Hockey IntelliGym. usahockeyintelligym.com/concussion-study-indicates-hockey-intelligym-training-could-reduce-chance-of-injury/.

51. "Science & Research." n.d. The Hockey IntelliGym. usahockeyintelligym.com/science-and-safety/; "Intelligym's Cognitive Therapy Technologies Added to Comprehensive Approach in Mayo Clinic Sports Medicine Center's Hockey Program." PRWeb. 2014. prweb.com/releases/2014/05/prweb11844861.htm; "Mayo Clinic Sports Medicine Center Adopts Intelligym Cognitive Training." SharpBrains. May 14, 2014. sharpbrains.com/blog/2014/05/14/mayo-clinic-sports-medicine-center-adopts-intelligym-cognitive-training/.

52. Morrison, Briana B., and Betsy DiSalvo. "Khan Academy Gamifies Computer Science."

Proceedings of the 45th ACM Technical Symposium on Computer Science Education—SIGCSE '14. 2014. doi.org/10.1145/2538862.2538946.

53. Rayner, Keith, et al. "So Much to Read, so Little Time." *Psychological Science in the Public Interest* 17 (1): 4–34, 2016. doi.org/10.1177/1529100615623267.

54. Adams, Tim. "Speed-Reading Apps: Can You Really Read a Novel in Your Lunch Hour?" *The Guardian.* April 8, 2017. theguardian.com/technology/2017/apr/08/speed-reading-apps-can-you-really-read-novel-in-your-lunch-hour; Rayner, et al., "So Much to Read, so Little Time"; pubmed.ncbi.nlm.nih.gov/26769745/; connection.ebscohost.com/c/articles/19236938/how-good-are-some-worlds-best-readers; Duggan, Geoffrey B., and Stephen J. Payne. "Skim Reading by Satisficing." *Proceedings of the 2011 Annual Conference on Human Factors in Computing Systems —CHI '11.* 2011. doi.org/10.1145/1978942.1979114.

55. Hutton, Elizabeth, and S. Shyam Sundar. "Can Video Games Enhance Creativity? Effects of Emotion Generated ByDance Dance Revolution." *Creativity Research Journal* 22 (3): 294–303, 2010. doi.org/10.1080/10400419.2010.503540.

56. "Want to Boost Creativity? Try Playing Minecraft." 2019. news.iastate.edu/news/2019/07/08/minecraftcreative; "Want to Boost Creativity? Try Playing Minecraft." ScienceDaily. 2019. sciencedaily.com/releases/2019/07/190708140051.htm.

Chapter 18: Zapping for the Better

1. "What Is Cranial Electrotherapy Stimulation (CES)?" n.d. Neuromodec.com. neuromodec.com/what-is-cranial-electrotherapy-stimulation-ces/.

2. "Transcutaneous Vagus Nerve Stimulation." n.d. PubMed. pubmed.ncbi.nlm.nih.gov/?term=%22transcutaneous+vagus+nerve+stimulation%22+%22cognition%22.

3. Cappabianca, Paolo, and Enrico de Divitiis. "Back to the Egyptians: Neurosurgery via the Nose. A Five-Thousand-Year History and the Recent Contribution of the Endoscope." *Neurosurgical Review* 30 (1): 1–7; discussion 7, 2007. doi.org/10.1007/s10143-006-0040-x.

4. Salehpour, Farzad, et al. "Brain Photobiomodulation Therapy: A Narrative Review." *Molecular Neurobiology* 55 (8): 6601–36, 2018. doi.org/10.1007/s12035-017-0852-4; "Brain Photobiomodulation." Vielight Inc. vielight.com/brain-photobiomodulation/; Zomorrodi, Reza, et al. "Pulsed near Infrared Transcranial and Intranasal Photobiomodulation Significantly Modulates Neural Oscillations: A Pilot Exploratory Study." *Scientific Reports* 9 (1), 2019. doi.org/10.1038/s41598-019-42693-x; Chao, Linda L. "Effects of Home Photobiomodulation Treatments on Cognitive and Behavioral Function, Cerebral Perfusion, and Resting-State Functional Connectivity in Patients with Dementia: A Pilot Trial." *Photobiomodulation, Photomedicine, and Laser Surgery* 37 (3): 133–41, 2019. doi.org/10.1089/photob.2018.4555; "Low-Level Laser Therapy." Wikipedia. en.wikipedia.org/wiki/Low-level_laser_therapy.

5. Medeiros, Liciane Fernandes, et al. "Neurobiological Effects of Transcranial Direct Current Stimulation: A Review." *Frontiers in Psychiatry* 3, 2012. doi.org/10.3389/fpsyt.2012.00110.

6. DaSilva, Alexandre F., et al. "Electrode Positioning and Montage in Transcranial Direct Current Stimulation." *Journal of Visualized Experiments* 51, 2011. doi.org/10.3791/2744.

7. Bikson, Marom, et al. "Safety of Transcranial Direct Current Stimulation: Evidence Based Update 2016." *Brain Stimulation* 9 (5): 641–61, 2016. doi.org/10.1016/j.brs.2016.06.004.

8. Kar, Kohitij, and Bart Krekelberg. "Transcranial Electrical Stimulation over Visual Cortex Evokes Phosphenes with a Retinal Origin." *Journal of Neurophysiology* 108 (8): 2173–78, 2012. doi.org/10.1152/jn.00505.2012.

9. Woods, A. J., et al. "A Technical Guide to TDCS, and Related Non-Invasive Brain Stimulation Tools." *Clinical Neurophysiology* 127 (2): 1031–48, 2016. doi.org/10.1016/j.clinph.2015.11.012.

10. Fields, R. Douglas. "Amping up Brain Function: Transcranial Stimulation Shows Promise in Speeding up Learning." *Scientific American.* November 25, 2011. scientificamerican.com/article /amping-up-brain-function/.

11. Monai, Hiromu, et al. "Calcium Imaging Reveals Glial Involvement in Transcranial Direct Current Stimulation-Induced Plasticity in Mouse Brain." *Nature Communications* 7 (1), 2016. doi.org/10.1038/ncomms11100.

12. Martin, Donel M., et al. "Can Transcranial Direct Current Stimulation Enhance Outcomes from Cognitive Training? A Randomized Controlled Trial in Healthy Participants." *International Journal of Neuropsychopharmacology* 16 (9): 1927–36, 2013. doi.org/10.1017 /s1461145713000539.

13. Nelson, Justin, et al. "The Effects of Transcranial Direct Current Stimulation (TDCS) on Multitasking Throughput Capacity." *Frontiers in Human Neuroscience* 10, 2016. doi.org/10.3389 /fnhum.2016.00589.

14. Kwon, Jung Won, et al. "The Effect of Transcranial Direct Current Stimulation on the Motor Suppression in Stop-Signal Task." *NeuroRehabilitation* 32 (1): 191–96, 2013. doi.org/10.3233 /NRE-130836.

15. Fregni, Felipe, et al. "Anodal Transcranial Direct Current Stimulation of Prefrontal Cortex Enhances Working Memory." *Experimental Brain Research* 166 (1): 23–30, 2005. doi.org /10.1007/s00221-005-2334-6.

16. Gill, Jay, Priyanka P. Shah-Basak, and Roy Hamilton. "It's the Thought That Counts: Examining the Task-Dependent Effects of Transcranial Direct Current Stimulation on Executive Function." *Brain Stimulation* 8 (2): 253–59, 2015. doi.org/10.1016/j.brs.2014.10.018.

17. Ditye, Thomas, et al. "Modulating Behavioral Inhibition by TDCS Combined with Cognitive Training." *Experimental Brain Research* 219 (3): 363–68, 2012. doi.org/10.1007/s00221 -012-3098-4.

18. Leadam, JD. n.d. "How I Zapped My Brain with a 9v Battery to Overcome Analysis Paralysis." Quantified Self. quantifiedself.com/show-and-tell/?project=891; Leadam, JD. n.d. "How I Shocked My Brain and Created Change." media.quantifiedself.com/slides/0891_JLeadam _HowZappedBrainWith9vBatteryOvercomeAnalysisParalysis.pdf.

18. Sandrini, Marco, et al. "Older Adults Get Episodic Memory Boosting from Noninvasive Stimulation of Prefrontal Cortex during Learning." *Neurobiology of Aging* 39: 210–16, 2016. doi .org/10.1016/j.neurobiolaging.2015.12.010.

20. Callaway, Ewen. "Shocks to the Brain Improve Mathematical Abilities." *Nature,* May, 2013. doi .org/10.1038/nature.2013.13012; Cohen Kadosh, Roi, et al. "Modulating Neuronal Activity Produces Specific and Long-Lasting Changes in Numerical Competence." *Current Biology* 20 (22): 2016–20, 2010. doi.org/10.1016/j.cub.2010.10.007.

21. Sarkar, A., A. Dowker, and R. Cohen Kadosh. "Cognitive Enhancement or Cognitive Cost: Trait-Specific Outcomes of Brain Stimulation in the Case of Mathematics Anxiety." *Journal of Neuroscience* 34 (50): 16605–10, 2014. doi.org/10.1523/jneurosci.3129-14.2014.

22. Matzen, Laura E., et al. "Effects of Non-Invasive Brain Stimulation on Associative Memory." *Brain Research* 1624: 286–96, 2015. doi.org/10.1016/j.brainres.2015.07.036.

23. Buch, Ethan R., et al. "Effects of TDCS on Motor Learning and Memory Formation: A Consensus and Critical Position Paper." *Clinical Neurophysiology: Official Journal of the International Federation of Clinical Neurophysiology* 128 (4): 589–603, 2017. doi.org/10.1016/j.clinph.2017.01.004.

24. "Depression, TDCS." Reddit. 2018. reddit.com/r/tDCS/search/?q=depression%2C%20tdcs &restrict_sr=1.

25. Kelley, Nicholas J., et al. "Stimulating Self-Regulation: A Review of Non-Invasive Brain

Stimulation Studies of Goal-Directed Behavior." *Frontiers in Behavioral Neuroscience* 12, 2019. doi.org/10.3389/fnbeh.2018.00337.

26. Ferrucci, R., et al. "Transcranial Direct Current Stimulation in Severe, Drug-Resistant Major Depression." *Journal of Affective Disorders* 118 (1-3): 215–19, 2009. doi.org/10.1016/j.jad.2009 .02.015; Fregni, Felipe, et al. "Treatment of Major Depression with Transcranial Direct Current Stimulation." *Bipolar Disorders* 8 (2): 203–4, 2006. doi.org/10.1111/j.1399-5618.2006.00291.x; Boggio, Paulo S., Soroush Zaghi, and Felipe Fregni. "Modulation of Emotions Associated with Images of Human Pain Using Anodal Transcranial Direct Current Stimulation (TDCS)." *Neuropsychologia* 47 (1): 212–17, 2009. doi.org/10.1016/j.neuropsychologia.2008.07.022; Kelley, "Stimulating Self-Regulation."

27. Sarkar, A., A. Dowker, and R. Cohen Kadosh. 2014. "Cognitive Enhancement."; Kelley, "Stimulating Self-Regulation."

28. Dambacher, Franziska, et al. "The Role of Right Prefrontal and Medial Cortex in Response Inhibition: Interfering with Action Restraint and Action Cancellation Using Transcranial Magnetic Brain Stimulation." *Journal of Cognitive Neuroscience* 26 (8): 1775–84, 2014. doi .org/10.1162/jocn_a_00595; Kelley, "Stimulating Self-Regulation."

29. Pripfl, Jürgen, and Claus Lamm. 2015. "Focused Transcranial Direct Current Stimulation (TDCS) over the Dorsolateral Prefrontal Cortex Modulates Specific Domains of Self-Regulation." *Neuroscience Research* 91 (February): 41–47. doi.org/10.1016/j.neures.2014.09.007; Kelley, "Stimulating Self-Regulation."

30. Hortensius, Ruud, Dennis J. L. G. Schutter, and Eddie Harmon-Jones. "When Anger Leads to Aggression: Induction of Relative Left Frontal Cortical Activity with Transcranial Direct Current Stimulation Increases the Anger–Aggression Relationship." *Social Cognitive and Affective Neuroscience* 7 (3): 342–47, 2011. doi.org/10.1093/scan/nsr012; Kelley, "Stimulating Self-Regulation."

31. Powers, Abigail, et al. "Effects of Combining a Brief Cognitive Intervention with Transcranial Direct Current Stimulation on Pain Tolerance: A Randomized Controlled Pilot Study." *Pain Medicine* 19 (4): 677–85, 2017. doi.org/10.1093/pm/pnx098; Kelley, "Stimulating Self-Regulation."

32. Nejati, Vahid, Mohammad Ali Salehinejad, and Michael A. Nitsche. "Interaction of the Left Dorsolateral Prefrontal Cortex (L-DLPFC) and Right Orbitofrontal Cortex (OFC) in Hot and Cold Executive Functions: Evidence from Transcranial Direct Current Stimulation (TDCS)." *Neuroscience* 369: 109–23, 2018. doi.org/10.1016/j.neuroscience.2017.10.042; Kelley, "Stimulating Self-Regulation."

33. Fregni, Felipe, et al. "Transcranial Direct Current Stimulation of the Prefrontal Cortex Modulates the Desire for Specific Foods." *Appetite* 51 (1): 34–41, 2008. doi.org/10.1016/j.appet .2007.09.016; Ljubisavljevic, M., et al. "Long-Term Effects of Repeated Prefrontal Cortex Transcranial Direct Current Stimulation (TDCS) on Food Craving in Normal and Overweight Young Adults." *Brain Stimulation* 9 (6): 826–33, 2016. doi.org/10.1016/j.brs.2016.07.002; Kelley, "Stimulating Self-Regulation."

34. Sacks, Oliver. "A Bolt from the Blue." *New Yorker.* July 16, 2007. newyorker.com/magazine /2007/07/23/a-bolt-from-the-blue.

35. Chi, Richard P., and Allan W. Snyder. "Facilitate Insight by Non-Invasive Brain Stimulation." *PLOS One* 6 (2): e16655, 2011. doi.org/10.1371/journal.pone.0016655.

36. Rosen, David S., et al. "Anodal TDCS to Right Dorsolateral Prefrontal Cortex Facilitates Performance for Novice Jazz Improvisers but Hinders Experts." *Frontiers in Human Neuroscience* 10, 2016. doi.org/10.3389/fnhum.2016.00579.

37. Horvath, Jared Cooney, Jason D. Forte, and Olivia Carter. "Quantitative Review Finds No Evidence of Cognitive Effects in Healthy Populations from Single-Session Transcranial Direct Current Stimulation (TDCS)." *Brain Stimulation* 8 (3): 535–50, 2015. doi.org/10.1016/j.brs.2015.01.400.

38. Wong, Lidia Y. X., Stephen J. Gray, and David A. Gallo. "Does TDCS over Prefrontal Cortex Improve Episodic Memory Retrieval? Potential Importance of Time of Day." *Cognitive Neuroscience* 9 (3-4): 167–80, 2018. doi.org/10.1080/17588928.2018.1504014.

39. Iuculano, T., and R. Cohen Kadosh. "The Mental Cost of Cognitive Enhancement." *Journal of Neuroscience* 33 (10): 4482–86, 2013. doi.org/10.1523/jneurosci.4927-12.2013.

40. "Roi Cohen Kadosh: Keeping Tabs on Transcranial Direct Current Stimulation." November 7, 2016. diytdcs.com/tag/roi-cohen-kadosh/.

41. Wurzman, Rachel, et al. "An Open Letter Concerning Do-It-Yourself Users of Transcranial Direct Current Stimulation." *Annals of Neurology* 80 (1): 1–4, 2016. doi.org/10.1002/ana.24689.

42. Nikolin, Stevan, et al. "Safety of Repeated Sessions of Transcranial Direct Current Stimulation: A Systematic Review." *Brain Stimulation* 11 (2): 278–88, 2018. doi.org/10.1016/j.brs.2017.10.020.

Chapter 19: A Pill a Day

1. "'I Learned Why They're Called Wonder Drugs: You Wonder What They'll Do to You.' Harlan Miller." Dictionary of Quotes. dictionary-quotes.com/i-learned-why-they-re-called-wonder-drugs-you-wonder-what-they-ll-do-to-you-harlan-miller/.

2. Giurgea, Corneliu. "Pharmacology of Integrative Activity of the Brain: Attempt at Nootropic Concept in Psychopharmacology" (in French). *Actualités Pharmacologiques* 25: 115–56, 1972. PMID 4541214.

3. Center for Food Safety and Applied Nutrition. "Peak Nootropics LLC Aka Advanced Nootropics—557887—02/05/2019." Center for Food Safety and Applied Nutrition. December 20, 2019. fda.gov/inspections-compliance-enforcement-and-criminal-investigations/warning-letters/peak-nootropics-llc-aka-advanced-nootropics-557887-02052019.

4. White, C. Michael. "Dietary Supplements Pose Real Dangers to Patients." *Annals of Pharmacotherapy* 54 (8): 815–19, 2020. doi.org/10.1177/1060028019900504.

5. "Arsenic, Lead Found in Popular Protein Supplements." *Consumer Reports.* March 12, 2018. consumerreports.org/dietary-supplements/heavy-metals-in-protein-supplements/.

6. White, "Dietary Supplements."

7. "FTC and FDA Send Warning Letters to Companies Selling Dietary Supplements Claiming to Treat Alzheimer's Disease and Remediate or Cure Other Serious Illnesses such as Parkinson's, Heart Disease, and Cancer." Federal Trade Commission. February 11, 2019. ftc.gov/news-events/press-releases/2019/02/ftc-fda-send-warning-letters-companies-selling-dietary.

8. "The Trouble with Mice as Behavioral Models for Alzheimer's." *STAT.* April 16, 2019. statnews.com/2019/04/16/trouble-mice-behavioral-models-alzheimers-neurologic-diseases/; Gribkoff, Valentin K., and Leonard K. Kaczmarek. "The Need for New Approaches in CNS Drug Discovery: Why Drugs Have Failed, and What Can Be Done to Improve Outcomes." *Neuropharmacology* 120 (July): 11–19, 2017. doi.org/10.1016/j.neuropharm.2016.03.021.

9. "Big Pharma Backed Away from Brain Drugs. Is a Return in Sight?" n.d. BioPharma Dive. biopharmadive.com/news/pharma-neuroscience-retreat-return-brain-drugs/570250/.

10. "Section 829." n.d. deadiversion.usdoj.gov. deadiversion.usdoj.gov/21cfr/21usc/829.htm.

11. "Don't Be Tempted to Use Expired Medicines." U.S. Food and Drug Administration, 2019. fda.gov/drugs/special-features/dont-be-tempted-use-expired-medicines.

12. "Prescription Stimulants." Drugabuse.gov. June 6, 2018. drugabuse.gov/publications/drugfacts/prescription-stimulants.

13. "Prescription Stimulants." Drugabuse.gov, June 6, 2018. drugabuse.gov/publications/drugfacts/prescription-stimulants.

14. repository.upenn.edu/neuroethics_pubs/130/; Ilieva, Irena P., Cayce J. Hook, and Martha J. Farah. "Prescription Stimulants' Effects on Healthy Inhibitory Control, Working Memory, and Episodic Memory: A Meta-Analysis." *Journal of Cognitive Neuroscience* 27 (6): 1069–89, 2015. doi.org/10.1162/jocn_a_00776; Bagot, Kara Simone, and Yifrah Kaminer. "Efficacy of Stimulants for Cognitive Enhancement in Non-Attention Deficit Hyperactivity Disorder Youth: A Systematic Review." *Addiction* 109 (4): 547–57, 2014. doi.org/10.1111/add.12460.

15. "Prescription Stimulants' Effects on Healthy Inhibitory Control, Working Memory, and Episodic Memory: A Meta-Analysis," 2015.

16. Ilieva, I., Boland, J., and Martha J. Farah. "Objective and Subjective Cognitive Enhancing Effects of Mixed Amphetamine Salts in Healthy People. *Neuropharmacology* 64, 496–505, 2013. doi: 10.1016/j.neuropharm.2012.07.021; Ilieva, Irena P., and Martha J. Farah. "Enhancement Stimulants: Perceived Motivational and Cognitive Advantages." *Frontiers in Neuroscience* 7, 2013. doi.org/10.3389/fnins.2013.00198.

17. Silver, Larry. "ADHD Neuroscience 101." ADDitude. November 30, 2006. additudemag.com/adhd-neuroscience-101/#:~:text=ADHD%20was%20the%20first%20disorder.

18. Tsai, Ching-Shu, et al. "Long-Term Effects of Stimulants on Neurocognitive Performance of Taiwanese Children with Attention-Deficit/Hyperactivity Disorder." *BMC Psychiatry* 13 (1), 2013. doi.org/10.1186/1471-244x-13-330.

19. "Prescription Stimulants." Drugabuse.gov.

20. "Brain Maturity Extends Well beyond Teen Years." NPR.org. October 10, 2011. npr.org/templates/story/story.php?storyId=141164708.

21. Acee, Anna M., and Leighsa Sharoff. "Herbal Remedies, Mood, and Cognition." *Holistic Nursing Practice* 26 (1): 38–51, 2021. doi.org/10.1097/hnp.0b013e31823bff70.

22. Bykov, Katsiaryna, PharmD. "CBD and Other Medications: Proceed with Caution." Harvard Health Blog, January 11, 2021. health.harvard.edu/blog/cbd-and-other-medications-proceed-with-caution-2021011121743.

23. Darbinyan, V., et al. "Rhodiola Rosea in Stress Induced Fatigue: A Double Blind Cross-over Study of a Standardized Extract SHR-5 with a Repeated Low-Dose Regimen on the Mental Performance of Healthy Physicians during Night Duty." *Phytomedicine: International Journal of Phytotherapy and Phytopharmacology* 7 (5): 365–71, 2000. doi.org/10.1016/S0944-7113(00)80055-0.

24. "Rhodiola." nccih.nih.gov/health/Rhodiola.

25. "RHODIOLA: User Ratings for Effectiveness, Side Effects, Safety and Interactions." n.d. Webmd. webmd.com/vitamins-supplements/ingredientreview-883-RHODIOLA.aspx?drugid=883&drugname=RHODIOLA.

26. Dimpfel, Wilfried, Leonie Schombert, and Alexander G. Panossian. "Assessing the Quality and Potential Efficacy of Commercial Extracts of Rhodiola Rosea L. by Analyzing the Salidroside and Rosavin Content and the Electrophysiological Activity in Hippocampal Long-Term Potentiation, a Synaptic Model of Memory." *Frontiers in Pharmacology* 9 (425): 1–11, 2018. doi.org/doi.org/10.3389/fphar.2018.00425.

27. Chandrasekhar, K., Jyoti Kapoor, and Sridhar Anishetty. "A Prospective, Randomized Double-Blind, Placebo-Controlled Study of Safety and Efficacy of a High-Concentration Full-Spectrum

Extract of Ashwagandha Root in Reducing Stress and Anxiety in Adults." *Indian Journal of Psychological Medicine* 34 (3): 255–62, 2012. doi.org/10.4103/0253-7176.106022.

28. "Withanolide: An Overview." n.d. ScienceDirect sciencedirect.com/topics/agricultural-and -biological-sciences/withanolide; Wal, Pranay, and Ankita Wal. "Chapter 34—an Overview of Adaptogens with a Special Emphasis on Withania and Rhodiola." In *Nutrition and Enhanced Sports Performance,* Academic Press, 343–50, 2013.

29. Salve, Jaysing, et al. "Adaptogenic and Anxiolytic Effects of Ashwagandha Root Extract in Healthy Adults: A Double-Blind, Randomized, Placebo-Controlled Clinical Study." *Cureus* 11 (12), 2019. doi.org/10.7759/cureus.6466.

30. Auddy, Biswajit, et al. "A Standardized Withania Somnifera Extract Significantly Reduces Stress-Related Parameters in Chronically Stressed Humans: A Double-Blind, Randomized, Placebo-Controlled Study." 2008. researchgate.net/publication/242151370_A_Standardized _Withania_Somnifera_Extract_Significantly_Reduces_Stress-Related_Parameters_in _Chronically_Stressed_Humans_A_Double-Blind_Randomized_Placebo-Controlled_Study.

31. Cooperman, Tod. n.d. "Ashwagandha Supplement Reviews & Information." ConsumerLab. consumerlab.com/reviews/ashwagandha-supplements/ashwagandha/#cautions.

32. Cooperman, "Ashwagandha Supplement Reviews & Information."

33. Sarris, Jerome, et al. "The Acute Effects of Kava and Oxazepam on Anxiety, Mood, Neurocognition; and Genetic Correlates: A Randomized, Placebo-Controlled, Double-Blind Study." *Human Psychopharmacology: Clinical and Experimental* 27 (3): 262–69, 2012. doi.org/10.1002 /hup.2216.

34. Volz, H.-P., and M. Kieser. "Kava-Kava Extract WS 1490 Versus Placebo in Anxiety Disorders: A Randomized Placebo-Controlled 25-Week Outpatient Trial." *Pharmacopsychiatry* 30 (1): 1–5, 1997. doi.org/10.1055/s-2007-979474.

35. Sarris, Jerome, Emma LaPorte, and Isaac Schweitzer. "Kava: A Comprehensive Review of Efficacy, Safety, and Psychopharmacology." *Australian & New Zealand Journal of Psychiatry* 45 (1): 27–35, 2011. doi.org/10.3109/00048674.2010.522554.

36. Scaccia, Annamarya. "Kava: Inside the All-Natural High That's Sweeping America." *Rolling Stone.* March 16; 2018. rollingstone.com/culture/culture-news/kava-inside-the-all-natural-high-thats -sweeping-america-125828/.

37. "Kava: A Review of the Safety of Traditional and Recreational Beverage Consumption Technical Report," n.d. fao.org/3/i5770e/i5770e.pdf.

38. Zuraw, Lydia. "NPR Choice Page." npr.org. 2019. npr.org/sections/thesalt/2013/04/24/178625554 /how-coffee-influenced-the-course-of-history.

39. O'Callaghan, Frances, Olav Muurlink, and Natasha Reid. "Effects of Caffeine on Sleep Quality and Daytime Functioning." *Risk Management and Healthcare Policy* 11 (December): 263–71, 2018. doi.org/10.2147/rmhp.s156404.

40. McLellan, Tom M., John A. Caldwell, and Harris R. Lieberman. "A Review of Caffeine's Effects on Cognitive, Physical and Occupational Performance." *Neuroscience & Biobehavioral Reviews* 71: 294–312, 2016. doi.org/10.1016/j.neubiorev.2016.09.001.

41. Smith, Andrew P. "Caffeine, Extraversion and Working Memory." *Journal of Psychopharmacology* 27 (1): 71–76, 2012. doi.org/10.1177/0269881112460111.

42. Haskell, Crystal F., et al. "The Effects of L-Theanine, Caffeine and Their Combination on Cognition and Mood." *Biological Psychology* 77 (2): 113–22, 2008. doi.org/10.1016/j.biopsycho .2007.09.008.

43. Camfield, David A., et al. "Acute Effects of Tea Constituents L-Theanine, Caffeine, and

Epigallocatechin Gallate on Cognitive Function and Mood: A Systematic Review and Meta-Analysis." *Nutrition Reviews* 72 (8): 507–22, 2014. doi.org/10.1111/nure.12120.

44. McCall, Rosie. "Yes, It's Possible to Have Too Much Caffeine (and These Are the Caffeine Overdose Symptoms to Look For)." Health.com. January 27, 2017. health.com/nutrition/caffeine-overdose-symptoms.

45. "Do I Need to Cycle Caffeine?" Examine.com. October 22, 2018. examine.com/nutrition/do-i-need-to-cycle-caffeine/.

46. Rogers, Peter J., et al. "Effects of Caffeine and Caffeine Withdrawal on Mood and Cognitive Performance Degraded by Sleep Restriction." *Psychopharmacology* 179 (4): 742–52, 2005. doi.org/10.1007/s00213-004-2097-y.

47. Dodd, F. L., et al. "A Double-Blind, Placebo-Controlled Study Evaluating the Effects of Caffeine and L-Theanine Both Alone and in Combination on Cerebral Blood Flow, Cognition and Mood." *Psychopharmacology* 232 (14): 2563–76, 2015. doi.org/10.1007/s00213-015-3895-0.

48. Holmgren, Per, Lotta Nordén-Pettersson, and Johan Ahlner. "Caffeine Fatalities—Four Case Reports." *Forensic Science International* 139 (1): 71–73, 2004. doi.org/10.1016/j.forsciint.2003.09.019.

49. Office of the Commissioner. "Spilling the Beans: How Much Caffeine Is Too Much?" FDA, September 2020. fda.gov/consumers/consumer-updates/spilling-beans-how-much-caffeine-too-much#:~:text=For%20healthy%20adults%2C%20the%20FDA.

50. 23andMe. "Caffeine Consumption & Genetics." 23andme.com/topics/wellness/caffeine-consumption/.

51. ncbi.nlm.nih.gov/pubmed/22747190; ncbi.nlm.nih.gov/pubmed/12404571; Pase, Matthew P., et al. "The Cognitive-Enhancing Effects of *Bacopa monnieri*: A Systematic Review of Randomized, Controlled Human Clinical Trials." *Journal of Alternative and Complementary Medicine* 18 (7): 647–52, 2012. doi.org/10.1089/acm.2011.0367; Nathan, P. J., et al. "The Acute Effects of an Extract of *Bacopa monniera* (Brahmi) on Cognitive Function in Healthy Normal Subjects." *Human Psychopharmacology: Clinical and Experimental* 16 (4): 345–51, 2001. doi.org/10.1002/hup.306.

52. Neale, Chris, et al. "Cognitive Effects of Two Nutraceuticals Ginseng and Bacopa Benchmarked against Modafinil: A Review and Comparison of Effect Sizes." *British Journal of Clinical Pharmacology* 75 (3): 728–37, 2013. doi.org/10.1111/bcp.12002.

53. Kongkeaw, Chuenjid, et al. "Meta-Analysis of Randomized Controlled Trials on Cognitive Effects of *Bacopa monnieri* Extract." *Journal of Ethnopharmacology* 151 (1): 528–35, 2014. doi.org/10.1016/j.jep.2013.11.008.

54. "Nootropics." 2013.

55. "Blood Alcohol Concentration." Rev. James E. McDonald, C.S.C., Center for Student Well-Being. Rev. 2019. mcwell.nd.edu/your-well-being/physical-well-being/alcohol/blood-alcohol-concentration/.

56. Jarosz, Andrew F., Gregory J. H. Colflesh, and Jennifer Wiley. "Uncorking the Muse: Alcohol Intoxication Facilitates Creative Problem Solving." *Consciousness and Cognition* 21 (1): 487–93, 2012. doi.org/10.1016/j.concog.2012.01.002.

57. Schuster, Julius, and Ellen S. Mitchell. "More than Just Caffeine: Psychopharmacology of Methylxanthine Interactions with Plant-Derived Phytochemicals." *Progress in Neuro-Psychopharmacology and Biological Psychiatry* 89: 263–74, 2019. doi.org/10.1016/j.pnpbp.2018.09.005.

58. Patterson, Brittany. "Can Tea Help Save the Amazon?" *Scientific American*, n.d. scientificamerican.com/article/can-tea-help-save-the-amazon/.

59. Schuster and Mitchell, "More than Just Caffeine"; Prediger, Rui D. S., et al. "Effects of Acute Administration of the Hydroalcoholic Extract of Mate Tea Leaves (Ilex Paraguariensis) in

Animal Models of Learning and Memory." *Journal of Ethnopharmacology* 120 (3): 465–73, 2008. doi.org/10.1016/j.jep.2008.09.018; Frank, Kurtis, et al. "Yerba Mate Research Analysis." Examine.com, September 2019. examine.com/supplements/yerba-mate/#; Frank, Kurtis, et al. "Lion's Mane Research Analysis." Examine.com, February 2021. examine.com/supplements /yamabushitake/#effect-matrix.

60. Harvard Health Publishing. "Nicotine: It May Have a Good Side." Harvard Health. March 9, 2014. health.harvard.edu/newsletter_article/Nicotine_It_may_have_a_good_side.

61. Heishman, Stephen J., Bethea A. Kleykamp, and Edward G. Singleton. "Meta-Analysis of the Acute Effects of Nicotine and Smoking on Human Performance." *Psychopharmacology* 210 (4): 453–69, 2010. doi.org/10.1007/s00213-010-1848-1.

62. Winblad, Bengt. "Piracetam: A Review of Pharmacological Properties and Clinical Uses." *CNS Drug Reviews* 11 (2): 169–82, 2006. doi.org/10.1111/j.1527-3458.2005.tb00268.x; Tariska, P., and Andras Paksy. "Cognitive Enhancement Effect of Piracetam in Patients with Mild Cognitive Impairment and Dementia." *Orvosi Hetilap* 141 (22): 1189–93, 2000; Wilsher, Colin R. "Effects of Piracetam on Developmental Dyslexia." *International Journal of Psychophysiology* 4 (1): 29–39, 1986. doi.org/10.1016/0167-8760(86)90048-6; Malykh, Andrei G., and M. Reza Sadaie. "Piracetam and Piracetam-like Drugs: From Basic Science to Novel Clinical Applications to CNS Disorders." *Drugs* 70 (3): 287–312, 2010. doi.org/10.2165/11319230 -000000000-00000.

63. Malykh and Sadaie, "Piracetam and Piracetam-like Drugs."

64. "Peak Nootropics LLC Aka Advanced Nootropics—557887—02/05/2019." Center for Food Safety and Applied Nutrition. December 20, 2019. fda.gov/inspections-compliance -enforcement-and-criminal-investigations/warning-letters/peak-nootropics-llc-aka-advanced- nootropics-557887-02052019.

65. Battleday, R. M., and A. K. Brem. "Modafinil for Cognitive Neuroenhancement in Healthy Non-Sleep-Deprived Subjects: A Systematic Review." *European Neuropsychopharmacology* 26 (2): 391, 2016. doi.org/10.1016/j.euroneuro.2015.12.023.

66. "Modafinil Side Effects: Common, Severe, Long Term." drugs.com/sfx/modafinil-side-effects .html.

67. "Can Microdosing Psychedelics Improve Your Mental Health?" Science in the News. December 18, 2020. sitn.hms.harvard.edu/flash/2020/can-microdosing-psychedelics-improve -your-mental-health/.

68. "QS Amsterdam 2017 Preview: A Year of Psilocybin Microdosing." Quantified Self. May 4, 2017. quantifiedself.com/blog/qs-amsterdam-2017-preview-year-psilocybin-micro-dosing/.

69. Feuer, Will. "Oregon Becomes First State to Legalize Magic Mushrooms as More States Ease Drug Laws in 'Psychedelic Renaissance.'" CNBC. November 4, 2020. cnbc.com/2020/11/04/ oregon-becomes-first-state-to-legalize-magic-mushrooms-as-more-states-ease-drug-laws .html.

70. sciencedirect.com/science/article/pii/S0924977X20309111; Hutten, Nadia R. P. W., et al. "Mood and Cognition after Administration of Low LSD Doses in Healthy Volunteers: A Placebo Controlled Dose-Effect Finding Study." *European Neuropsychopharmacology* 41: 81–91, 2020. doi.org/10.1016/j.euroneuro.2020.10.002; biologicalpsychiatryjournal.com/article /S0006-3223(19)31409-X/fulltext; Bershad, Anya K., et al. "Acute Subjective and Behavioral Effects of Microdoses of Lysergic Acid Diethylamide in Healthy Human Volunteers." *Biological Psychiatry* 86 (10): 792–800, 2019. doi.org/10.1016/j.biopsych.2019.05.019.

Chapter 20: Sci Fi to Sci Fact

1. As of a check in January 2021, 23andMe didn't offer cognition-related interpretations anymore.
2. Allegrini, A. G., et al. "Genomic Prediction of Cognitive Traits in Childhood and Adolescence." *Molecular Psychiatry* 24 (6): 819–27, 2019. doi.org/10.1038/s41380-019-0394-4.
3. Payton, Antony. "The Impact of Genetic Research on Our Understanding of Normal Cognitive Ageing: 1995 to 2009." *Neuropsychology Review* 19 (4): 451–77, 2009. doi.org/10.1007/s11065-009-9116-z.
4. Ericsson, Anders. *Peak*. New York: Vintage. 2017.
5. "The California Prenatal Screening Program." cdph.ca.gov. California Department of Public Health, Genetic Disease Screening Program. March 2017. cdph.ca.gov/Programs/CFH/DGDS/CDPH%20Document%20Library/PNS%20Documents/Patient%20Booklet%20Consent_ENG-ADA.pdf; Porter, Forbes D. "Smith–Lemli–Opitz Syndrome: Pathogenesis, Diagnosis and Management." *European Journal of Human Genetics* 16 (5): 535–41, 2008. doi.org/10.1038/ejhg.2008.10.
6. Araki, Motoko, and Tetsuya Ishii. "International Regulatory Landscape and Integration of Corrective Genome Editing into in Vitro Fertilization." *Reproductive Biology and Endocrinology* 12 (1): 108, 2014. doi.org/10.1186/1477-7827-12-108.
7. Hollingsworth, Julia, and Isaac Yee. "Chinese Authorities: 1st Gene-Edited Babies Were Illegal." CNN. December 30, 2019. cnn.com/2019/12/30/china/gene-scientist-china-intl-hnk/index.html.
8. Araki and Ishii, "International Regulatory Landscape."
9. Stein, Rob. "House Committee Votes to Continue Ban on Genetically Modified Babies." National Public Radio. June 4, 2019. npr.org/sections/health-shots/2019/06/04/729606539/house-committee-votes-to-continue-research-ban-on-genetically-modified-babies.
10. Powell, Alvin. "CRISPR's Breakthrough Implications." *Harvard Gazette*. May 16, 2018. news.harvard.edu/gazette/story/2018/05/crispr-pioneer-jennifer-doudna-explains-gene-editing-technology-in-prather-lectures/#:~:text=The%20Cas9%20protein%20(short%20for.
11. "The Science and Engineering of Intelligence: A Bridge across Vassar Street." The Center for Brains, Minds & Machines. Massachusetts Institute of Technology. January 15, 2016. cbmm.mit.edu/science-engineering-vassar; Zhang, Feng. "Improving, Applying and Extending CRISPR-Cas9 Systems." Presented at the Science and Engineering of Intelligence: A Bridge across Vassar Street, January 15, 2016.
12. "DIY Bacterial Gene Engineering CRISPR Kit." n.d. The ODIN. the-odin.com/diy-crispr-kit/.
13. Taub, Eric A. "Sleepy behind the Wheel? Some Cars Can Tell." *New York Times*. March 16, 2017. nytimes.com/2017/03/16/automobiles/wheels/drowsy-driving-technology.html.
14. Perlow, Jon. "New in Labs: Stop Sending Mail You Later Regret." Official Gmail Blog. Google, October 6, 2008. gmail.googleblog.com/2008/10/new-in-labs-stop-sending-mail-you-later.html.
15. Iawama, Gabrie, et al. "Introducing the Decision Advisor: A Simple Online Tool That Helps People Overcome Cognitive Biases and Experience Less Regret in Real-Life Decisions." Paper presented at the 40th Annual Meeting of the Society for Judgement and Decision Making, Montreal, November 15–18, 2019. doi.org/10.13140/RG.2.2.10816.07689.
16. Chui, Michael, James Manyika, and Mehdi Miremadi. "Where Machines Could Replace Humans—and Where They Can't (Yet)." *McKinsey Quarterly*. McKinsey & Company. July 8, 2016. mckinsey.com/business-functions/mckinsey-digital/our-insights/where-machines-could-replace-humans-and-where-they-cant-yet.
17. Savage, Maddy. "Thousands of Swedes Are Inserting Microchips under Their Skin." National

Public Radio. October 22, 2018. npr.org/2018/10/22/658808705/thousands-of-swedes-are -inserting-microchips-under-their-skin.

18. Coffey, Helen. "This Swedish Rail Company Is Letting Commuters Pay Using Microchips in Their Hands." *Independent*. June 16, 2017. independent.co.uk/travel/news-and-advice/sj-rail -train-tickets-hand-implant-microchip-biometric-sweden-a7793641.html.

19. "XEM RFID Chip." n.d. Dangerous Things. dangerousthings.com/product/xem/.

20. Rodriguez, Salvador. "Facebook Agrees to Acquire Brain-Computing Start-up CTRL-labs." CNBC. September 23, 2019. cnbc.com/2019/09/23/facebook-announces-acquisition-of-brain -computing-start-up-ctrl-labs.html.

21. "F8 2017: AI, Building 8 and More Technology Updates from Day Two." Facebook. April 19, 2017. about.fb.com/news/2017/04/f8-2017-day-2/.

22. Eveleth, Rose. "Google Glass Wasn't a Failure. It Raised Crucial Concerns." *Wired*. December 12, 2018. wired.com/story/google-glass-reasonable-expectation-of-privacy/; Levy, Karyne. "A Surprising Number of Places Have Banned Google Glass in San Francisco." *Business Insider*. March 18, 2014. businessinsider.com/google-glass-ban-san-francisco-20143.

23. Eagleman, David. 2015. "Can We Create New Senses for Humans?" TED. 2015. ted.com/talks /david_eagleman_can_we_create_new_senses_for_humans?language=en.

24. "Neosensory Announces $10 Million Series a Financing Round." *Markets Insider*. January 9, 2019. markets.businessinsider.com/news/stocks/neosensory-announces-10-million-series-a -financing-round-1027855882.

25. "Infrared Vision." *National Geographic*. October 17, 2013. nationalgeographic.org/media /infrared-vision/.

26. Rosen, Rebecca J. "6 Animals That Can See or Glow in Ultraviolet Light." *The Atlantic*. August 15, 2011. theatlantic.com/technology/archive/2011/08/6-animals-that-can-see-or-glow-in-ultraviolet -light/243634/.

27. Keller, Kate. "Could This Futuristic Vest Give Us a Sixth Sense?" *Smithsonian Magazine*. April 20, 2018. smithsonianmag.com/innovation/could-this-futuristic-vest-give-us-sixth-sense-180968852/; King, Darryn. "Hearing Loss? A New Device Lets You Feel Sound." *Wall Street Journal*. September 5, 2019. wsj.com/articles/hearing-loss-a-new-device-lets-you-feel-sound-11567691822; Kotler, Steven. "Neuroscientist David Eagleman Aims to Give Deaf People a New Way to Hear—and Upgrade Everyone Else's Senses Too." NEO.LIFE. January 10, 2019. neo. life/2019/01/the-wristband-that-gives-you-superpowers/; Walters, Helen. "How to Hear the World through Your Back." TED Conferences. March 17, 2015. ideas.ted.com/how-to -hear-the-world-through-your-chest/.

28. Eberhardt, Jennifer. *Biased*. New York: Penguin Books. 2020.

29. Watts, Alexander W. "Why Does John Get the STEM Job rather than Jennifer?" The Clayman Institute for Gender Research. Stanford University. June 2, 2014. gender.stanford.edu/news -publications/gender-news/why-does-john-get-stem-job-rather-jennifer.

30. Vartan, Starre. "Racial Bias Found in a Major Health Care Risk Algorithm." *Scientific American*. October 24, 2019. scientificamerican.com/article/racial-bias-found-in-a-major-health-care-risk -algorithm/.

31. Reuters. "Amazon Ditched AI Recruiting Tool That Favored Men for Technical Jobs." *The Guardian*. October 11, 2018. theguardian.com/technology/2018/oct/10/amazon-hiring-ai-gender -bias-recruiting-engine.

32. Larson, Jeff, et al. "How We Analyzed the COMPAS Recidivism Algorithm." ProPublica. May 23, 2016. propublica.org/article/how-we-analyzed-the-compas-recidivism-algorithm.

33. Markoff, John. "Elon Musk's Neuralink Wants 'Sewing Machine-Like' Robots to Wire Brains

to the Internet." *New York Times.* July 17, 2019. nytimes.com/2019/07/16/technology/neuralink-elon-musk.html.

34. Markoff, "Elon Musk's Neuralink"; Rogers, Adam. "Here's How Elon Musk Plans to Put a Computer in Your Brain." *Wired.* July 17, 2019. wired.com/story/heres-how-elon-musk-plans-to-stitch-a-computer-into-your-brain/.

35. Sanders, Robert. "Sprinkling of Neural Dust Opens Door to Electroceuticals." *Berkeley News.* August 3, 2016. news.berkeley.edu/2016/08/03/sprinkling-of-neural-dust-opens-door-to-electroceuticals/.

36. Sanders, "Sprinkling of Neural Dust"; Marr, Bernard. "Smart Dust Is Coming. Are You Ready?" *Forbes.* September 16, 2018. forbes.com/sites/bernardmarr/2018/09/16/smart-dust-is-coming-are-you-ready/#5e426a3f5e41.

37. Sutcliffe, Magdalena, and Madeline A. Lancaster. "A Simple Method of Generating 3D Brain Organoids Using Standard Laboratory Equipment." *Methods in Molecular Biology* 1576: 1–12, 2017. doi.org/10.1007/7651_2017_2.

38. Yeager, Ashley. "Infographic: How to Make a Brain Organoid." *The Scientist.* July 31, 2018. the-scientist.com/infographics/infographic--how-to-make-a-brain-organoid-64534; "Center for Genomically Engineered Organs." n.d. Harvard University. cgeo.hms.harvard.edu/; "Mark Skylar-Scott." n.d. Stanford Bioengineering. Stanford University. bioengineering.stanford.edu/people/mark-skylar-scott.

39. Whiteman, Honor. "Your Brain Could Be Backed Up, for a Deadly Price." medicalnewstoday.com. Healthline Media UK, March 16, 2018. medicalnewstoday.com/articles/321235#How-does-vitrifixation-work?.

40. Letzter, Rafi. "MIT Just Cut Ties with Nectome, the '100-Percent-Fatal' Brain-Preserving Company." Live Science. April 2, 2018. livescience.com/62202-mit-nectome-brain-upload.html.

41. Bell, Gordon, and Jim Gray. "Digital Immortality." *Communications of the ACM* 44 (3): 28–31, 2001. doi.org/10.1145/365181.365182.

42. Hamilton, Isobel Asher. "2 Tech Founders Lost Their Friends in Tragic Accidents. Now They've Built AI Chatbots to Give People Life after Death." Insider. November 17, 2018. businessinsider.com/eternime-and-replika-giving-life-to-the-dead-with-new-technology-2018-11; Öhman, Carl, and Luciano Floridi. "An Ethical Framework for the Digital Afterlife Industry." *Nature Human Behaviour* 2 (5): 318–20, 2018. doi.org/10.1038/s41562-018-0335-2.

43. Eckstein, Maria K., et al. "Beyond Eye Gaze: What Else Can Eyetracking Reveal about Cognition and Cognitive Development?" *Developmental Cognitive Neuroscience* 25 (June 2017): 69–91. doi.org/10.1016/j.dcn.2016.11.001.

44. "Detect and Monitor Brain Diseases through Typing Cadence." neurametrix.com/blog/2019/01/21/fda-grants-breakthrough-device-designation-to-neurametrix-101.

45. Johnson, Bryan. "A Plan for Humanity." Medium. December 31, 2019. medium.com/future-literacy/a-plan-for-humanity-2bc04088e3d4.

46. "GDPR.eu." 2019. gdpr.eu/.

47. Johnson, "A Plan for Humanity."

48. Johnson, "A Plan for Humanity."

49. There are many, but here are a few: Upwork (general), Catalant (businesspeople and domain experts), Toptal (engineers), Dribbble (designers), Kolabtree (for scientists), Reedsy (editors).

50. "Achievement." myachievement.com/about.

51. Yong, Ed. "Psychology's Replication Crisis Is Running Out of Excuses." *The Atlantic.* November 19, 2018. theatlantic.com/science/archive/2018/11/psychologys-replication-crisis-real/576223/.

52. "MIT Solve | Sapien Labs." solve.mit.edu/challenges/brain-health/solutions/96.

53. "MIT Solve | Sapien Labs."

54. "About Us." Sapien Labs. sapienlabs.org/about-us/.

55. "About Us." Sapien Labs, n.d. sapienlabs.org/about-us/.

Chapter 21: Your 15-Minute Self-Experiments

1. McGlothlin, Anna E., and Roger J. Lewis. "Minimal Clinically Important Difference." *JAMA* 312 (13): 1342, 2014. doi.org/10.1001/jama.2014.13128; Wright, Alexis, et al. "Clinimetrics Corner: A Closer Look at the Minimal Clinically Important Difference (MCID)." *Journal of Manual & Manipulative Therapy* 20 (3): 160–66, 2012. doi.org/10.1179/20426186 12y.0000000001.

2. Rozenkrantz, Liron, et al. "Placebo Can Enhance Creativity." Edited by Emmanuel Manalo. *PLOS One* 12 (9): e0182466, 2017. doi.org/10.1371/journal.pone.0182466.

3. Kovacevic, Ana, et al. "The Effects of Aerobic Exercise Intensity on Memory in Older Adults." *Applied Physiology, Nutrition, and Metabolism,* October 2019. doi.org/10.1139/apnm-2019 -0495; Oppezzo, Marily, and Daniel L. Schwartz. "Give Your Ideas Some Legs: The Positive Effect of Walking on Creative Thinking." *Journal of Experimental Psychology: Learning, Memory, and Cognition* 40, 2014; Stenfors, Cecilia U. D., et al. "Positive Effects of Nature on Cognitive Performance Across Multiple Experiments: Test Order but Not Affect Modulates the Cognitive Effects." *Frontiers in Psychology* 10, July 2019. doi.org/10.3389/fpsyg.2019.01413; Steinberg, H., et al. "Exercise Enhances Creativity Independently of Mood." *British Journal of Sports Medicine* 31 (3): 240–45, 1997. doi.org/10.1136/bjsm.31.3.240.

4. Lifehack. 2013. "7-Minute Workout." YouTube. youtube.com/watch?v=ECxYJcnvyMw; Reynolds, Gretchen. "The Scientific 7-Minute Workout." *New York Times,* May 9, 2013. well.blogs .nytimes.com/2013/05/09/the-scientific-7-minute-workout/.

5. Iowa State University News Service. "Want to Boost Creativity? Try Playing Minecraft." Iastate .edu, 2019. news.iastate.edu/news/2019/07/08/minecraftcreative; Hutton, Elizabeth, and S. Shyam Sundar. "Can Video Games Enhance Creativity? Effects of Emotion Generated by Dance Dance Revolution." *Creativity Research Journal* 22 (3): 294–303, 2010. doi.org/10.1080 /10400419.2010.503540.

6. "Get Minecraft." Minecraft.net, September 29, 2020. minecraft.net/en-us/get-minecraft/.

7. Chrysikou, Evangelia G., et al. "Noninvasive Transcranial Direct Current Stimulation over the Left Prefrontal Cortex Facilitates Cognitive Flexibility in Tool Use." *Cognitive Neuroscience* 4 (2): 81–89, 2013. doi.org/10.1080/17588928.2013.768221.

8. Chrysikou, et al., "Noninvasive Transcranial Direct Current Stimulation."

9. Mineo, Liz. "Less Stress, Clearer Thoughts with Mindfulness Meditation." *Harvard Gazette,* April 17, 2018. news.harvard.edu/gazette/story/2018/04/less-stress-clearer-thoughts-with-mindfulness -meditation/.

10. Flett, Jayde A. M., et al. "Mobile Mindfulness Meditation: A Randomised Controlled Trial of the Effect of Two Popular Apps on Mental Health." *Mindfulness* 10 (5): 863–76, 2018. doi .org/10.1007/s12671-018-1050-9; Mani, Madhavan, et al. "Review and Evaluation of Mindfulness-Based IPhone Apps." *JMIR MHealth and UHealth* 3 (3): e82, 2015. doi.org/10.2196 /mhealth.4328.

11. Guevarra, Darwin A., et al. "Placebos Without Deception Reduce Self-Report and Neural Measures of Emotional Distress." *Nature Communications* 11 (1): 3785, 2020. doi.org/10.1038 /s41467-020-17654-y; Schaefer, Michael, et al. "Open-Label Placebos Reduce Test Anxiety and

Improve Self-Management Skills: A Randomized-Controlled Trial." *Scientific Reports* 9 (1), 2019. doi.org/10.1038/s41598-019-49466-6.

12. Spiridon, Elena, and Stephen Fairclough. "The Effects of Ambient Blue Light on Anger Levels: Applications in the Design of Unmanned Aircraft GCS." *International Journal of Unmanned Systems Engineering* 5 (3): 53–69, 2017. doi.org/10.14323/ijuseng.2017.8; De Kort, Yvonne, et al. "Lighting and Self-Regulation: Can Light Revitalize the Depleted Ego?" 2017.

13. Iyadurai, L., et al. "Preventing Intrusive Memories After Trauma via a Brief Intervention Involving Tetris Computer Game Play in the Emergency Department: A Proof-of-Concept Randomized Controlled Trial." *Molecular Psychiatry* 23 (3): 674–82, 2017. doi.org/10.1038/mp.2017.23; Lobel, Adam, et al. "Designing and Utilizing Biofeedback Games for Emotion Regulation." *Proceedings of the 2016 CHI Conference Extended Abstracts on Human Factors in Computing Systems*, May 2016. doi.org/10.1145/2851581.2892521; Bakker, David, et al. "Mental Health Smartphone Apps: Review and Evidence-Based Recommendations for Future Developments." *JMIR Mental Health* 3 (1): e7, 2016. doi.org/10.2196/mental.4984; Payne, Hannah E., Victor B. A. Moxley, and Elizabeth MacDonald. "Health Behavior Theory in Physical Activity Game Apps: A Content Analysis." *JMIR Serious Games* 3 (2): e4, 2015. doi.org/10.2196/games.4187; Roepke, Ann Marie, et al. "Randomized Controlled Trial of SuperBetter, a Smartphone-Based/Internet-Based Self-Help Tool to Reduce Depressive Symptoms." *Games for Health Journal* 4 (3): 235–46, 2015. doi.org/10.1089/g4h.2014.0046; Devan, Hemakumar, et al. "Evaluation of Self-Management Support Functions in Apps for People with Persistent Pain: Systematic Review." *JMIR MHealth and UHealth* 7 (2): e13080, 2019. doi.org/10.2196/13080.

14. Sarris, Jerome, Emma LaPorte, and Isaac Schweitzer. "Kava: A Comprehensive Review of Efficacy, Safety, and Psychopharmacology." *Australian and New Zealand Journal of Psychiatry* 45 (1): 27–35, 2011. doi.org/10.3109/00048674.2010.522554.

15. Lover, Kava. "Genuine Kava—Helping Your Search for Certified Roots Kava.com." The Trusted Kava Source | Kava Root Powder | Kava Wholesale, March 23, 2017. kava.com/genuine-kava/.

16. "Menu." Kava Lounge SF, 2016. kavaloungesf.com/menu/.

17. "Buy Fine Kava | Kava Powders | Instant Kava | Bottled Kava | Green Kava | the Kava Society." Kava Society, 2011. kavasociety.nz/shop.

18. "Menu." Kava Lounge SF.

19. Chowhound. "Kava Shake Recipe." Chowhound, n.d. chowhound.com/recipes/kava-shake-14098.

20. Adam, Hajo, and Adam D. Galinsky. "Enclothed Cognition." *Journal of Experimental Social Psychology* 48 (4): 918–25, 2012. doi.org/10.1016/j.jesp.2012.02.008.

21. Hsieh, Shu-Shih, et al. "Systematic Review of the Acute and Chronic Effects of High-Intensity Interval Training on Executive Function across the Lifespan." *Journal of Sports Sciences* 39 (1): 10–22, 2020. doi.org/10.1080/02640414.2020.1803630; Stenfors, Cecilia U. D., et al. "Positive Effects of Nature on Cognitive Performance Across Multiple Experiments: Test Order but Not Affect Modulates the Cognitive Effects." *Frontiers in Psychology* 10 (July 2019). doi.org/10.3389/fpsyg.2019.01413; Hsieh, Shu-Shih, et al. "Systematic Review of the Acute and Chronic Effects of High-Intensity Interval Training on Executive Function across the Lifespan." *Journal of Sports Sciences* 39 (1): 10–22, 2020. doi.org/10.1080/02640414.2020.1803630.

22. Stenfors, Cecilia U. D., et al. "Positive Effects of Nature."

23. "10 Minute Taekwondo Workout." YouTube, n.d. youtube.com/watch?v=Ujly7l3-xUM.

24. "Surya Namaskar." Wikipedia, November 9, 2020. en.wikipedia.org/wiki/Surya_Namaskar.

25. Beaven, C. Martyn, and Johan Ekström. "A Comparison of Blue Light and Caffeine Effects on Cognitive Function and Alertness in Humans." Edited by Denis Burdakov. *PLOS One* 8 (10): e76707, 2013. doi.org/10.1371/journal.pone.0076707.

26. Bhayee, Sheffy, et al. "Attentional and Affective Consequences of Technology Supported Mindfulness Training: A Randomised, Active Control, Efficacy Trial." *BMC Psychology* 4 (November 2016). doi.org/10.1186/s40359-016-0168-6.

27. Smith, Glenn E., et al. "A Cognitive Training Program Based on Principles of Brain Plasticity: Results from the Improvement in Memory with Plasticity-Based Adaptive Cognitive Training (IMPACT) Study." *Journal of the American Geriatrics Society* 57 (4): 594–603, 2009. doi .org/10.1111/j.1532-5415.2008.02167.x; Jaeggi, S. M., et al. "Improving Fluid Intelligence with Training on Working Memory." *Proceedings of the National Academy of Sciences* 105 (19): 6829– 33, 2008. doi.org/10.1073/pnas.0801268105; Smith, Glenn E., et al. "A Cognitive Training Program Based on Principles of Brain Plasticity: Results from the Improvement in Memory with Plasticity-Based Adaptive Cognitive Training (IMPACT) Study." *Journal of the American Geriatrics Society* 57 (4): 594–603, 2009. doi.org/10.1111/j.1532-5415.2008.02167.x.

28. Imburgio, Michael J., and Joseph M. Orr. "Effects of Prefrontal TDCS on Executive Function: Methodological Considerations Revealed by Meta-Analysis." *Neuropsychologia* 117 (August 2018): 156–66. doi.org/10.1016/j.neuropsychologia.2018.04.022.

29. Imburgio and Orr, "Effects of Prefrontal TDCS."

30. Haskell, Crystal F., et al. "The Effects of L-Theanine, Caffeine and Their Combination on Cognition and Mood." *Biological Psychology* 77 (2): 113–22, 2008. doi.org/10.1016/j.biopsycho .2007.09.008.

31. Denis, M. 1985. "Visual Imagery and the Use of Mental Practice in the Development of Motor Skills." *Canadian Journal of Applied Sport Sciences. Journal canadien des sciences appliquées au sport* 10 (4): 4S16S. pubmed.ncbi.nlm.nih.gov/3910301/; Pascual-Leone, A., et al. "Modulation of Muscle Responses Evoked by Transcranial Magnetic Stimulation during the Acquisition of New Fine Motor Skills." *Journal of Neurophysiology* 74 (3): 1037–45, 1995. doi.org/10.1152 /jn.1995.74.3.1037; Kaptchuk, Ted J., et al. "Placebos Without Deception: A Randomized Controlled Trial in Irritable Bowel Syndrome." Edited by Isabelle Boutron. *PLOS One* 5 (12): e15591, 2010. https://doi.org/10.1371/journal.pone.0015591.

32. Winter, Bernward, et al. "High Impact Running Improves Learning." *Neurobiology of Learning and Memory* 87 (4): 597–609, 2007. doi.org/10.1016/j.nlm.2006.11.003.

33. Marzbani, H., H. Marateb, and M. Mansourian. "Methodological Note: Neurofeedback: A Comprehensive Review on System Design, Methodology and Clinical Applications." *Basic and Clinical Neuroscience Journal* 7 (2), 2016. doi.org/10.15412/j.bcn.03070208.

34. Karpicke, J. D., and J. R. Blunt. "Retrieval Practice Produces More Learning than Elaborative Studying with Concept Mapping." *Science* 331 (6018): 772–75, 2011. doi.org/10.1126 /science.1199327.

35. Neale, Chris, et al. "Cognitive Effects of Two Nutraceuticals Ginseng and Bacopa Benchmarked against Modafinil: A Review and Comparison of Effect Sizes." *British Journal of Clinical Pharmacology* 75 (3): 728–37, 2013. doi.org/10.1111/bcp.12002.

36. "Beginners—Nootropics." Reddit, 2013.

37. Frank, Kurtis, et al. "*Bacopa monnieri* Research Analysis." Examine.com, February 2019. exam ine.com/supplements/bacopa-monnieri/.

38. Pase, Matthew P., et al. "The Cognitive-Enhancing Effects of *Bacopa monnieri*: A Systematic Review of Randomized, Controlled Human Clinical Trials." *Journal of Alternative and*

Complementary Medicine 18 (7): 647–52, 2012. doi.org/10.1089/acm.2011.0367; Nathan, P. J., et al. "The Acute Effects of an Extract of *Bacopa monnieri* (Brahmi) on Cognitive Function in Healthy Normal Subjects." *Human Psychopharmacology: Clinical and Experimental* 16 (4): 345–51, 2001. doi.org/10.1002/hup.306.

Chapter 22: Pretty Pictures

1. Schel, M. A., and T. Klingberg. "Specialization of the Right Intraparietal Sulcus for Processing Mathematics During Development." *Cerebral Cortex* 27 (9): 4436–46, 2017. doi.org/10.1093/cercor/bhw246. PMID: 27566976; en.wikipedia.org/wiki/Intraparietal_sulcus#/media/File:Gray726_intraparietal_sulcus.svg; health.qld.gov.au/abios/asp/boccipital.
2. sciencedirect.com/science/article/abs/pii/S1053811905024663?dgcid=api_sd_search-api-endpoint; news.mit.edu/2014/in-the-blink-of-an-eye-0116.
3. Olive, Melissa L., and Jessica H. Franco. "(Effect) Size Matters: And So Does the Calculation." *Behavior Analyst Today* 9 (1): 5–10, 2008. doi.org/10.1037/h0100642.
4. If these charts looked familiar, that's because they are my own take on Anscombe's Quartet, a famous statistics data set posed by English statistician Francis Anscombe to illustrate a similar point about the dangers of using averages without actually looking at data graphically.

Final Goodies, Goodbye, and Good Luck!

1. Othmer, Siegfried. Email to Elizabeth Ricker. March 15, 2021.

Index

About the Author

ELIZABETH R. RICKER has been self-experimenting and studying neurohacking for the last 10 years. Her work has been featured on public broadcast TV in Belgium and in the March for Science's book *Science Not Silence* (MIT Press, 2018). Ricker is a sought-after expert by Silicon Valley venture capital firms, technology startups, and Fortune 500 companies, and she has given talks on cognitive enhancement around the US and in China. Previously, she held research, product, and strategy roles in healthcare and educational technology startups that served over 70 million users. Ricker conducted neuroscience and educational research at MIT, the Media Lab, and Harvard. She holds an undergraduate degree in Brain and Cognitive Sciences from MIT and a graduate degree in Mind, Brain, and Education from Harvard. Find out what she's up to in neuroscience, tech, and other adventures (as well as more content related to this book) at ericker.com.